GAME THEORY, SOCIAL CHOICE AND ETHICS

Edited by
H. W. BROCK

Reprinted from:
Theory and Decision, Vol. 11, Nos. 2/3

D. REIDEL PUBLISHING COMPANY

DORDRECHT : HOLLAND / BOSTON : U.S.A.
LONDON : ENGLAND

ISBN-13: 978-94-009-9534-5   e-ISBN-13978-94-009-9532-1
DOI: 10.1007/978-94-009-9532-1

TABLE OF CONTENTS

FOREWORD

There are problems to whose solution I would attach an infinitely greater importance than to those of mathematics, for example touching ethics, or our relation to God, or concerning our destiny and our future; but their solution lies wholly beyond us and completely outside the province of science.

J. F. C. Gauss

For all his prescience in matters physical and mathematical, the great Gauss apparently did not foresee one development peculiar to our own time. The development I have in mind is the use of mathematical reasoning – in particular the axiomatic method – to explicate alternative concepts of rationality and morality. The present bipartite collection of essays (Vol. 11, Nos. 2 and 3 of this journal) is entitled 'Game Theory, Social Choice, and Ethics'. The eight papers represent state-of-the-art research in formal moral theory. Their intended aim is to demonstrate how the methods of game theory, decision theory, and axiomatic social choice theory can help to illuminate ethical questions central not only to moral theory, but also to normative public policy analysis.

Before discussion of the contents of the papers, it should prove helpful to recall a number of pioneering papers that appeared during the decade of the 1950s. These papers contained a series of mathematical and conceptual breakthrough which laid the basis for much of today's research in formal moral theory. The papers deal with two somewhat distinct topics: the concept of individual and collective rationality, and the concept of social justice. Let me first discuss the important advances that were made in our understanding of rational behavior.

Building on the seminal work of von Neumann and Morgenstern (1947), J. Marschak (1950) set forth a clear and accessible account of the theory of individual decision-making under conditions of 'risk'. Specifically, Marschak

Theory and Decision **11** (1979) 143–151.   0040–5833/79/0112–0143$00.90

derived the expected utility maximization theorem from a streamlined set of axioms. At about the same time, L. J. Savage (1954) undertook an investigation of utility theory which focused not only on the meaning of rational behavior (i.e. expected utility maximization), but also on the concept of 'subjective probability' and its relation to the concepts of preference and rationality.

Simultaneously, Savage was joined by Hurwicz (1951), Chernoff (1954), Milnor (1954) and others who analyzed the difficult problem of characterizing rational behavior under conditions of 'complete ignorance' (as opposed to 'risk'). By the mid-1950s therefore, an extensive literature had appeared dealing with the problem of individually rational behavior under conditions of risk and uncertainty.

At the level of group decision-making, research proceeded in two rather different directions. First there was the work of Arrow (1951) in axiomatic social choice theory. Arrow investigated the sense in which it was meaningful to extend the concept of individual rationality under conditions of certainty from the context of an individual person to that of a collectivity. In doing so, he arrived at his celebrated impossibility theorem.

Second, there was the work of game theorists who addressed the problem of multi-person decision-making in a variety of ways, all quite different from that of Arrow. Nash (1950) introduced the concept of an equilibrium point of a non-cooperative game, and proved that every finite non-cooperative game of complete information possesses at least one equilibrium point. Nash (1951) also introduced a careful distinction between 'cooperative' and 'non-cooperative' games. Finally, he (Nash, 1950a, 1953) axiomatized a unique solution to non-zero-sum two-person cooperative games. This is the well-known Nash 'bargaining solution'.

Whereas Nash arrived at his bargaining solution by analyzing properties that a reasonable outcome of bargaining might have, Harsanyi (1956, 1961) arrived at the same solution concept by another route. He introduced a set of postulates concerning rational behavior *and* rational expectations that parties can entertain about one another during the bargaining process and showed that these postulates lead to the Nash solution.

In the province of *n*-person game theory, Shapley (1953) proved that every *n*-person game with side-payments will possess a unique solution satisfying a set of abstract axioms pertinent to games in characteristic function form.

This solution has been called the 'Shapley Value' of a game. Harsanyi (1959, 1963) subsequently generalized Nash's 1953 two-person bargaining model to the n-person case, and showed that his bargaining solution coincides with a generalization of the Shapley Value to the non-side-payments case. At the same time as Shapley's original work, Gillies (1953) introduced the notion of the core of an n-person cooperative game, and Milnor (1952) characterized a set of cooperative game outcomes that he called 'reasonable'.

The problem of social justice received less attention than the problem of rational behavior during the 1950s. However, the work that was done was of considerable importance. Harsanyi (1953) deduced the utilitarian sum-of-the-utilities rule from a simple model of 'impersonal' rational decision. Specifically, a citizen is said to be impersonally situated if he faces an equal chance of ending up in the shoes of any of n citizens when a policy decision must be taken. If an impersonally situated citizen is rational, it follows from the expected utility maximization theorem that he will select that policy which maximizes the arithmetic mean of the total utility accruing to all n citizens, or equivalently which maximizes the expected utility associated with his own 'extended' utility function. We thus have an elegant linkage between the modern theory of rational choice under risk and classical utilitarianism.

Fleming (1952) and Harsanyi (1955) each gave social choice theoretic axiomizations of the utilitarian welfare function. Fleming's paper makes use of postulates that refer to individual and social preferences between sure prospects only. Harsanyi altered Fleming's argument somewhat, and generalized it to the case of uncertain prospects. Specifically, he required that both individual and social preferences satisfy the Marschak postulates for rational choice under risk. His 1955 result provided an additional source of support for his earlier model of impersonal moral choice.

Raiffa (1953) and Braithwaite (1955) both introduced game theoretically inspired 'arbitration schemes' for two-person cooperative game problems. Shapley showed that his Value for n-person cooperative games awards each player a utility payoff equal to his 'average' marginal contribution of utility to the various coalitions he might join in random order. Shapley's result showed that the Value incorporates a concept of 'strategic equity', and can serve as a model for the age-old concept of distribution according to relative contribution.

Rawls (1958) confronted the broad problem of social justice and introduced his concept of 'justice as fairness'. His account of justice in this early paper made use of a hypothetical bargaining game. However, subsequent accounts (e.g., Rawls, 1971) of his theory replace the bargaining game with a Veil of Ignorance construct similar to Harsanyi's model of impersonal rational decision, although there are significant differences among the details of the two theories. Finally, Harsanyi (1958) clarified the metaethical status of moral rules by arguing that such rules qualify as hypothetical imperatives, rather than as Kantian categorical imperatives.

The collection of essays presented here stands as a testament to the enduring influence of these earlier works on current research in moral theory. The authors make systematic use of the analytical perspectives that were in large measure the legacy of the 1950s: Arrowian social choice theory; cooperative and non-cooperative game theory; decision theory under conditions both of 'risk' and 'complete ignorance'. The following tableau displays the use of these various theories by author.

But if the present essays reflect this past research, they also mirror the important advances of more recent years. Two recent developments in moral theory are of particular significance, and are salient in several of the ensuing papers. First, ongoing research has appreciably deepened our understanding

	E	SCT	DT-R	DT-U	GT
Barry	•				•
Brock	•	•			•
Gibbard	•	•			
Harsanyi	•	•	•		
Maskin	•	•		•	
Schwartz	•	•			
Strasnick	•	•			
Wittman	•	•	•	•	•

Legend:
  E — Ethics
  SCT — Axiomatic social choice theory
  DT-R — Decision theory in the case of "risk"
  DT-U — Decision theory in the case of "complete ignorance"
  GT — Game theory

of the delicate interplay between substantive ethical theory on the one hand, and axiomatic theories of individual and collective rationality on the other hand. For example, the moral concept of allocation according to relative need can be explicated by means of axiomatic cooperative game theory *and* Arrowian social choice theory *and* decision theory. The work by Kaneko, Nakamura, Strasnick, Harsanyi and others which has made this advance possible is critically reviewed in my own paper.

In a second important development, our understanding of alternative theories of ethics and rational choice has been enhanced by a systematic investigation of the symmetry conditions and invariance relationships which characterize these theories. Recent articles by d'Aspremont and Gevers (forthcoming) and by Sen (1977) summarize much of this research. And the present papers by Harsanyi, Maskin, Strasnick, and Wittman contribute to our understanding of the problem. It should be stressed that this problem is not merely technical in nature, nor need it be addressed in a starkly formal manner. For what is ultimately at stake here is a set of substantive ethical questions such as: Does/should a given ethical theory depend upon a 'zero point'? Does/should the theory entail interpersonal comparisons of utility, and if so, what kind of comparisons? And more generally, what kind of information is admissible in a given moral argument?

Let me now briefly describe the contents of the eight essays. I have rather arbitrarily divided the papers into two groups of four, reflecting the fact that the essays fill the pages of two consecutive issues of *Theory and Decision*. The papers will be discussed in the order in which they appear in this double issue.

Barry reviews R. B. Braithwaite's insightful dilemma of the pianist and the trumpeter who occupy adjoining rooms. He first provides a critical discussion of several ethical and game theoretical solutions to Braithwaite's problem. Thereafter he introduces a new solution and deduces its implications for three variations of the dilemma. Finally, he investigates the implications of his solution for broader issues of distributive justice. Schwartz argues that long-range welfare policies cannot be justified by an appeal to the welfare of remote future generations. By 'long-range welfare policies' Schwartz simply means policies which provide significant benefits to future generations at some cost to present generations. His conclusion is based on the observation that the failure to adopt such a policy would not make any of our descendents worse off than he would otherwise be, since, had the policy been adopted, *he* would not even have *existed*. Schwartz's argument brings out some interesting conflicts between utilitarian and Paretian principles.

Strasnick addresses the conflict between the Rawlsian and the utilitarian theories of justice. He argues that the two theories need not be at odds with one another. For each is appropriate in a particular class of moral situations. The first part of his paper discusses the general structure of moral theory and argues that morality need not be identified with any single moral principle. A framework useful for analyzing issues of distributive justice is introduced in the second part of the paper. Then in the final portion of the essay, this framework is applied to the analysis of two different moral situations. Strasnick argues that the utilitarian principle is appropriate for the first situation – one in which a scarce good must be efficiently distributed. And he claims that Rawls' Difference Principle is the correct one for the second situation in which the more abstract issue of basic institutional justice arises. Wittman shows how several different analytical theories of justice can be analyzed within the context of a uniform diagrammatic exposition. His essay accordingly provides an elegant and accessible overview of various theories of justice. Additionally he sets forth numerous propositions which establish some of the critical differences between alternative theories, and the conditions under which different theories yield identical results.

My own paper discusses a new theory of justice I have developed in the past few years. The theory is Contractarian in spirit, and makes use of the Nash-Harsanyi-Shapley theory of the valuation of n-person cooperative games. The theory integrates into a coherant whole two fundamental distributive norms: To Each according to his Needs; and To Each according to his Contribution. The theory also incorporates a new account of ethics in terms of impartial decision – an account which dispenses with the need for a Veil of Ignorance construct. Finally, my theory makes possible a new interpretation of two cooperative game solution concepts: the Nash solution and the generalized Shapley Value. Gibbard attempts to clarify Rawls' theory of 'primary goods' by means of social choice theoretic reasoning. Rawls' Difference Principle asserts that a basic economic structure is just if it makes the worst off people as well off as possible. How well off someone is can supposedly be measured by an index of primary social goods. Rawls himself has given no clear account of the way in which such an index can be constructed. Gibbard proposes a new version of the Difference Principle which invokes a partial ordering of prospects for opportunities, rather than an index of primary goods.

Harsanyi reexamines the logical foundations of Bayesian decision theory in the first part of his paper, and argues that the Bayesian criterion of expected utility maximization is the only decision criterion consistent with rationality. He then reviews his classical results dating from the 1950s which show that the Bayesian criterion in conjunction with the Paretian criterion entail a utilitarian theory of morality. Next he discusses the role both of cardinal utility and of (cardinal) interpersonal comparisons in ethics. He shows that by making use of information not available in Arrowian social choice theory, the utilitarian welfare function satisfies analogues of all of Arrow's original conditions. Finally, Harsanyi contrasts rule and act utilitarianism, and suggests why the former is preferable for the purposes of ethical theory. In the final essay, Maskin launches a systematic new investigation into the problem of decision-making in the face of 'complete ignorance', and links his result to the theory of social choice. In the first section of his paper, Maskin introduces a set of properties which might characterize a criterion for decision-making under complete ignorance. Two of these properties are novel in the present context: 'independence of non-discriminating states', and 'weak pessimism'. The second section provides a new characterization of the principle of insufficient reason. In the third part, lexicographic maximin and maximax criteria are characterized. Finally, Maskin links his results to the problem of social choice.

This collection of essays originated in a request Charles Plott made of me to organize three panels of speakers for the 1978 Annual Public Choice Society Meetings in New Orleans. I am indebted both to Charles Plott and to Kenneth Arrow for helpful suggestions as to paper topics and participants for these panels on Game Theory, Social Choice, and Ethics.

Menlo Park, California   HORACE W. BROCK

BIBLIOGRAPHY

Arrow, K. J.: 1951, Social Choice and Individual Values, Cowles Commission Monograph 12 (John Wiley & Sons, New York).

Aspremont, C. de and L. Gevers: forthcoming, 'Equity and the informational basis of collective choice', Review of Economic Studies.

Braithwaite, R. B.: 1955, Theory of Games as a Tool for the Moral Philosopher (Cambridge University Press, Cambridge).

Chernoff, Herman: 1954, 'Rational selection of decision functions', Econometrica 22, pp. 422–443.

Fleming, Marcus: 1952, 'A cardinal concept of welfare', Quarterly Journal of Economics, pp. 366–384.

Gillies, D. B.: 1953, Some Theoresm on n-Person Games, Ph.D. thesis, Department of Mathematics (Princeton University, Princeton).

Harsanyi, John C.: 1953, 'Cardinal utility in welfare economics and in the theory of risk-taking', Journal of Political Economy 61.

Harsanyi, John C.: 1955, 'Cardinal welfare, individualistic ethics, and interpersonal comparisons of utility', Journal of Political Economy 63.

Harsanyi, John C.: 1956, 'Approaches to the bargaining problem before and after the theory of games: A critical discussion of Zeuthen's, Hick's, and Nash's theories', Econometrica 24, pp. 144–157.

Harsanyi, John C.: 1958, 'Ethics in terms of hypothetical imperatives', Mind 47, pp. 305–316.

Harsanyi, John C.: 1959, 'A bargaining model for the cooperative n-person game', Annals of Mathematics Studies, No. 40.

Harsanyi, John C.: 1961, 'On the rationality postulates underlying the theory of cooperative games', Journal of Conflict Resolution 5.

Harsanyi, John C.: 1963, 'A simplified bargaining model for the n-personal cooperative game', International Economic Review 4.

Hurwicz, Leonid: 1951, Optimality Criteria for Decision Making Under Ignorance, Cowles Commission Discussion Paper, Statistics, No. 370.

Marschak, Jacob: 1950, 'Rational behavior, uncertain prospects, and measurable utility', Econometrica 18, pp. 111–141.

Milnor, J. W.: 1952, 'Reasonable outcomes for n-person games', Research Memorandum RM-916 (The RAND Corporation, Santa Monica).

Milnor, J. W.: 1954, 'Games against nature', in Thrall, Coombs, and Davis (eds.), pp. 49–60.

Nash, J. F.: 1950, 'Equilibrium points in n-person games', Proceedings of the National Academy of Sciences, U.S.A. 36, pp. 48–49.

Nash, J. F.: 1950a, 'The bargaining problem', Econometrica 18, pp. 155–162.

Nash, J. F.: 1951, 'Non-cooperative games', Annals of Mathematics 54, pp. 286–295.

Nash, J. F.: 1953, 'Two-person cooperative games', Econometrica 21, pp. 128–140.

Raiffa, Howard: 1953, 'Arbitration schemes for generalized two-person games', Annals of Mathematics Studies, No. 28.

Rawls, John: 1958, 'Justice as fairness', Philosophy Review 67, No. 164.

Rawls, John: 1971, A Theory of Justice (The Bellknap Press of Harvard University, Cambridge).

Savage, L. J.: 1945, The Foundations of Statistics (John Wiley & Sons, New York, and Chapman & Hall, London).

Sen, Amartya K.: 1977, 'On weights and measures: Informational constraints in social welfare analysis', Econometrica, Vol. 45.

Shapley, Lloyd S.: 1953, 'A value for n-person games', Annals of Mathematics Studies, No. 28.
Von Neumann, John and Oskar Morgenstern: 1947, Theory of Games and Economic Behavior, Second Edition (Princeton University Press, Princeton).

BRIAN BARRY

DON'T SHOOT THE TRUMPETER –
HE'S DOING HIS BEST!

Reflections on a Problem of Fair Division

ABSTRACT. R. B. Braithwaite's story of the pianist and trumpeter with adjoining rooms is presented, and his solution to the problem of dividing playing time fairly between them is discussed, along with alternative solutions that have been put forward. These solutions are criticized for the shared assumption that the object of distribution is utility rather than opportunity. The author proposes a different approach and works out its implications for the problem of the pianist and the trumpeter in three variants. In conclusion, the relevance of the analysis to wider issues of distribution is briefly considered.

In his Inaugural Lecture as Knightbridge Professor of Moral Philosophy in the University of Cambridge, R. B. Braithwaite posed the following problem:

Suppose that Luke and Matthew are both bachelors, and occupy flats in a house which has been converted into two flats by an architect who had ignored all considerations of acoustics. Suppose that Luke can hear everything louder than a conversation that takes place in Matthew's flat, and vice versa; but that sounds in the two flats do not penetrate outside the house. Suppose that it is legally impossible for either to prevent the other from making as much noise as he wishes, and economically or sociologically impossible for either to move elsewhere. Suppose further that each of them has only the hour from 9 to 10 in the evening for recreation, and that it is impossible for either to change to another time. Suppose that Luke's form of recreation is to play classical music on the piano for an hour at a time, and that Matthew's amusement is to improvise jazz on the trumpet for an hour at once. And suppose that whether or not either of them performs on one evening has no influence, one way or the other, upon the desires of either of them to perform on any other evening; so that each evening's happenings can be treated independently. Suppose that the satisfaction each derives from playing his instrument for the hour is affected, one way or the other, by whether or not the other is also playing: in radio language, there is 'interference' between them, positive or negative. Suppose that they put to me the problem: Can any plausible principle be devised stating how they should divide the proportion of days on which both of them play, Luke alone plays, Matthew alone plays, neither play, so as to obtain maximum production of satisfaction compatible with fair distribution? (Braithwaite, 1955, pp. 8–9.)

The story, as told by Braithwaite, comes with a payoff matrix representing the utilities that Luke and Matthew derive from each of the possible things that can happen on any evening between nine and ten. This is shown in Table I. The first number in each pair of payoffs represents Luke's utility, the second

Theory and Decision 11 (1979) 153–180.   0040–5833/79/0112–0153$02.80

TABLE I

| | | Matthew | |
		Play	Not play
Luke	*Play*	(1, 2)	(7, 3)
	Not play	(4, 10)	(2, 1)

Matthew's. There are, of course, a variety of ways of interpreting these numbers, and solutions differ in the amount of information that they need to extract. But taking the utilities in their weakest interpretation, as expressing the ordering of the four possible outcomes for the protagonists, we can see the essential structure of the problem. Each would most prefer to play alone, and each would next most prefer to have the other play alone. (This unusual second preference – hearing the other play being preferred by each to silence – should be borne in mind throughout.) The third and fourth preferences differ: Luke prefers silence to cacophony while Matthew's order is the reverse. This difference – widely referred to as Matthew's 'threat advantage' – has been made much of in the literature on the Luke and Matthew problem. (Too much, I shall argue.) It is clear that both silence and cacophony are inefficient. That is to say, for any outcome consisting of neither playing or both playing, there exists at least one outcome that both men would prefer. As it happens, there are a wide range of outcomes that both would prefer if we consider a sequence of evenings: Luke playing solo all the time, Matthew playing solo all the time, and any combination of Luke solo and Matthew solo. All of these possibilities are efficient in that whenever one of them is realized a move to any other would entail a gain for one of the men but a loss for the other. Thus, the Paretian frontier consists of all possible ways of dividing up the time so that on any given evening either Matthew is playing and Luke is listening or Luke is playing and Matthew is listening. A common perspective, which is shared by Braithwaite and others who have proposed solutions, is that the problem consists of finding a point on the efficiency frontier that is 'fair' in the sense that it shares equitably between Matthew and Luke the advantage that they can jointly achieve by cooperation.

In the paper which I began by quoting, Braithwaite (1955) proposed his own solution. This solution has been discussed by Lucas (1959), in an article devoted entirely to the problem; by Luce and Raiffa (1957, pp. 145–50) extensively and by Sen (1970, pp. 121–3) more briefly in their synthetic works; and in a half-page footnote by Rawls (1971, pp. 134–5, n. 10). Alternative ways of resolving the conflict between Matthew and Luke have been put forward by Lucas (1959, pp. 9–10); by Luce and Raiffa (1957, pp. 145–8), in accordance with a solution originally proposed by Raiffa; and by Gauthier (1974b, p. 64). The Nash solution (Nash, 1950; see also Luce and Raiffa, 1957, pp. 123–4 and 140–3; and Harsanyi, 1977, pp. 141–66) can also be applied to the case of Matthew and Luke. In this paper I shall review the criticisms of and alternatives to Braithwaite's solution that have been offered and offer my own solution. This solution is Rawlsian in spirit though Rawls is not, of course, to be held responsible for the use made here of his ideas.

Although the proposed solutions are all in some sense normative, they put the problem in different contexts – arbitration by a moral philosopher, agreement on general principles of distribution to be reached in advance (perhaps in a Rawlsian original position), and bargaining in the actual case by perfectly 'rational' agents. At the same time, we should not overlook the existence of a strong family resemblance among the solutions, which brings them into strong contrast with other conceptions of fairness, such as maximizing utility, equalizing utility, maximizing the minimum utility, dividing up the time equally, and so on. What they have in common is that (for whatever reason) they seek to make the cooperative outcome somehow reflect the payoffs available to the players in the absence of cooperation. More specifically, each solution consists of two components: a rule for determining the non-cooperative baseline and a rule for getting from that to some point on the Pareto frontier. I shall look at the disagreements among those who share this general approach before turning to criticisms directed at the approach itself.

As far as the baseline is concerned, two contenders for the position have found support. The first is that we should take the non-cooperative outcome to be what happens if Luke and Matthew each tries to inflict the maximum damage on the other. Inspection of the payoff matrix shows that this entails both Matthew and Luke playing. It may be noticed that this is not in fact the worst thing that could happen to Matthew: silence would be even worse for him. But within the framework of the story (in which Luke cannot spoil Matthew's *embouchure* with a punch in the head, smash his trumpet, etc.) there is no way in which Luke can by any action of his produce an outcome of silence. If he refrains from playing he simply provides Matthew with the chance for a solo performance. This proposal for establishing a baseline is the one proposed by Braithwaite himself, and also employed by Luce and Raiffa in the presentation of Raiffa's solution.

Leaving aside any ethical objections there may be to this way of establishing the baseline, it is not at all clear that self-interested maximizers would necessarily be well-advised to start by seeing how much damage each can inflict on the other. Suppose that cacophony gives Luke a severe migraine lasting several days, so that (on the same scale of utilities as Braithwaite's payoffs) an evening of cacophony scores minus a thousand. If Matthew knows this, Luke's threat to play is, looking at things in a common-sense way, not worth much. (If Matthew does not know it, Luke may be able to bluff him into thinking he doesn't mind cacophony much, but that gets us into questions of strategic manipulation that fall beyond the scope of the bargaining theories we are considering here.) In terms of the mathematics of the theories of fair division under discussion, if the baseline is two units for Matthew and minus a thousand for Luke, the other figures being as given by Braithwaite, Luke does very badly in the cooperative solution anyway. In general, there is no reason to suppose that the optimum threat is the same as the maximum threat. If I can injure you to a certain degree at the cost of my life, and a little less at much lower cost to me, it seems reasonable to say that I would be well advised to make the lesser threat.[2]

However, if what we are looking for is a fair division, it is also relevant to ask whether the fair outcome ought, as an ethical matter, to depend on the amount of damage each of the protagonists can inflict on the other. (It may be noted that this objection would lie as much against optimal threats as against maximal threats.) This complaint was voiced by Braithwaite's first critic, J. R. Lucas, and has been echoed by Sen and Rawls, all of whom pick up the phrase that Matthew has the 'threat advantage' because he prefers cacophony to silence whereas cacophony is Luke's most disliked outcome. I shall show later that Sen and Rawls really object not just to threats but to any intrusion of considerations of bargaining advantage. Lucas, however, remains faithful to the basic idea of Braithwaite's analysis — that the fair solution is one that divides the gains from cooperation so as to preserve the relative advantages of the two men under a non-cooperative regime.

His counterproposal is, therefore, that we set the baseline at the point where Matthew and Luke are doing their best rather than their worst. "Luke would pursue the prudential strategy, the one designed to secure his own interests, and would not depart from this merely to retaliate on Matthew [and vice versa]" (Lucas, 1959, p. 9). The question is, of course, what 'doing their best' (i.e. doing the best for themselves) means in this context. Obviously it does not mean 'obtaining the maximum feasible payoff', since the whole point of the problem is supposed to be that, in the absence of an arbitrated solution or an agreement, both do less well than they might. In practice, Lucas picks as baseline the security level of each player. That is to say, he takes the maximin mixed strategy for each player (Luke plays one night in four, Matthew one night in five) and treats as baseline values the payoffs that are generated by both men playing their maximin strategies.[3] David Gauthier, in his analysis of the case of Luke and Matthew, also picks the joint maximin as baseline (Gauthier, 1974b, p. 64).

The obvious objection to this is that there is no general reason why somebody who is trying to do as well for himself as possible should play a maximin strategy unless he believes he is engaged in a zero-sum game. And presumably Matthew and Luke are aware that they are not playing a zero-sum game. Otherwise the setting for the problem — the potential for mutual gain — disappears. Neither Gauthier nor Lucas seems to me to provide an adequate rationale for the choice of the joint maximin.

Gauthier says (1974b, p. 64) that the baseline is each player's 'minimal utility', which is "the worst he can do, whatever the circumstances and actions of the others; this is his maximin utility". But why equate playing a maximin strategy with doing the best for oneself? Suppose Matthew knows that Luke is playing his maximin strategy: he knows, let's say, that at nine every evening Luke flips two pennies and then plays the piano for an hour if and only if both pennies come up heads. Matthew's obviously best move is to wait until just after nine each evening, and adapt his behavior to Luke's. If he does this, playing whenever he has ascertained that Luke will not play, he can get 10 units on an average of three nights in four, and 3 units on an average of one night in four. This yields an expected utility of 8.25 units as against the 2.8 which is all that he would get in this situation by playing his own maximin strategy.[4] Even if, for some reason, he has to decide whether to play or not without waiting to see what Luke does, the information that Luke is playing his maximin strategy should lead him to depart from his own.

He is much better off with a strategy of playing every night (average utility 7.75) than with the 2.8 that his maximin strategy would yield.

The trouble is, of course, that Luke is symmetrically placed in this respect. If he can count on Matthew pursuing *his* maximin strategy, he can improve on the utility derivable from playing his own maximin strategy (3.25) by playing only when Matthew does not (6.4) or, if he has to choose in advance, playing every evening (5.8). But if *each* starts acting on the expectation that the other will pursue a maximin strategy, they are both disappointed. If they both decide in advance on a strategy rather than wait and see what the other does each evening, they both play all the time, so we are back with the original baseline to which joint maximin was supposed to be a superior alternative.

Thus, any situation in which each plays his maximin strategy is highly unstable, because it always pays to depart from one's own maximin strategy provided the other does not. We could still, if we wished, specify that the joint maximin payoffs are to be taken as baseline values as the first step in the construction of a 'fair' procedure; but the question is, why should we? Lucas suggests that the joint maximin is what would come about "if neither party realized that the musical interference was attributable to a person, and each assumed it was a natural phenomenon". For, under these beliefs,

each would pursue a prudential policy; on discovering that it was a colleague who was making the noise, and in coming to an agreement with him to remedy matters the *status quo ante* would be a natural starting point for arranging how matters should be improved. (Lucas, 1959, p. 10.)

But, if Matthew believed that the noises that sometimes emanate from the wall between nine o'clock and ten o'clock in the evening were natural phenomena, he would have *less*, not *more*, reason to adopt a maximin strategy. (As we have seen, it is not evident that a maximin strategy is rational if he thinks he is playing a non-zero-sum game with a human being. But it is definitely not rational against 'nature'.) In fact, the way to maximize his satisfaction if he thinks nothing he does will alter the frequency of sounds in the wall is simply to wait each evening and see if the sounds occur or not. If they do, he listens; if not, he plays. Supposing that for some reason he has to commit himself in advance to a strategy (though why should he?) it should be not a maximin strategy – in fact not a mixed strategy at all – but a decision either to play every evening or never to play. (Which he should choose depends on the estimate he makes of the future frequency of the mysterious noises.)

It is perhaps a tribute to the persuasive force of Braithwaite's paper that the two philosophers who have rejected the baseline constituted by Matthew and Luke each gratuitously inflicting the maximum possible injury on one another should nevertheless take as baseline a position in which Matthew and Luke are still inflicting injury on one another, though only incidentally to the pursuit of their maximin strategies. Surely there is an obvious alternative that one might take as the baseline, namely the outcome where neither plays.

In the case as stated by Braithwaite, of course, silence is ranked below either playing or listening by both men and is well below the security level for each. One might be led by that to say that the baseline should not be something that neither would choose. But suppose we change the case so that Luke has little interest in either playing or listening: he most wants silence, and if he can't have it he'd prefer hearing Matthew through the wall to adding to the din himself. (It does not make any difference where he puts playing solo himself, so long as he ranks it below silence.) And suppose Matthew prefers to play alone but is prepared to endure cacophony rather than abandon his trumpet. His preference order is: himself play, both play, Luke play, neither play. If, as Gauthier and Lucas propose, the baseline is established by each doing the best for himself without regard to the other, Matthew will finish up playing every evening and Luke (reluctantly) listening. For, whatever Luke does, Matthew is better off playing; and, whatever Matthew does, Luke is better off not playing. There are in this case no gains to be had from cooperation because any move away from the outcome where each is independently doing the best for himself must make Matthew worse off. By playing solo every evening he is already getting what he most wants. If Luke appeals for an adjudication to provide a fair division of evenings between what Matthew wants (to play the trumpet) and what he wants (silence), he will have to be told that the outcome is already Pareto-optimal so there is no room for 'fair division' to enter in. (Notice that Luke is weakened by being condemned to maximize: if he were allowed to create an inefficiency by gratuitously playing himself, even when he would prefer to listen, he could set the stage for a 'fair division'.)

The question is, of course, whether the fact that one of the people involved *might* prefer silence to either playing or listening is sufficient reason to set the baseline at silence. I think that it is. Whether or not Luke actually prefers listening to Matthew to silence, he should have a *right* to enjoy silence, and the same for Matthew. The baseline ought to be set at the point where each man is using his rights to prevent activity by the other of the kind that some people (even if not they) find objectionable. This is after all the way affairs are normally conducted: I have a right against my neighbor to stop his (say) putting up a structure that will block my light, because many people would not wish to have their light blocked; and my neighbor still has to get permission to build it even if he happens to know that I am an eccentric who keeps the blinds down all day and would actually prefer to have him build.

It might be said against this that *a priori* (in other words without knowing anything about the particular preferences of Matthew and Luke) silence and cacophony are both symmetrical arrangements. Braithwaite could therefore argue that if his proposal, to make cacophony the baseline, arbitrarily favors Matthew (who prefers cacophony to silence), mine equally arbitrarily favors Luke (who likes silence better). But the choice is arbitrary only if one believes that there is no basic difference between being let alone and being affected by another. I am of course perfectly well aware of all the difficulties that inhere in this distinction. The critical literature on J. S. Mill's distinction between self-regarding and other-regarding actions enforces that point strongly. All the same, I do believe that, in the case of Matthew and Luke, playing an instrument is an other-regarding act while things that they might do without producing effects perceived by the other would be self-regarding acts. I would therefore hold that, although silence and cacophony are both symmetrical, we can pick out silence as the preferred baseline on the strength of the general principle of not affecting others in ways that might be adverse.

Lucas recommends his baseline as what would be appropriate if

Dr. Matthew and Professor Luke are fellows of one of those colleges where the fellows are not friends but not enemies either, and treat one another in a distant though not discourteous fashion. (Lucas, 1959, p. 10.)

I agree that it would be inappropriate for them to make life unpleasant for one another merely to gain a bargaining advantage. But would it not be most appropriate for each of them not to make a noise that the other might find disagreeable unless they reach an agreement otherwise? Colleges do, in fact, have such rules; and even if the college of Dr. Matthew and Professor Luke does not, that is no reason why a moral philosopher called in to adjudicate should not take as a status quo point what would happen if such a rule were enforced.

I shall not in this paper go into the technical details of the way in which the various theories of fair division derive their answers to the question of how many nights Luke should play and how many nights Matthew should play. It is enough for the present purpose to bear in mind that all of them involve some procedure for moving from the baseline (that is, whichever baseline is regarded as appropriate) to the Pareto frontier.

All except the Nash solution have the notion of preserving equal relative advantage. Graphically, this entails transforming the utilities in some way to make them comparable, and then drawing a 45° line from the point representing the baseline payoffs up and to the right until it cuts the Pareto frontier. Braithwaite, Raiffa and Gauthier all have different ways of transforming the utilities. A method which does not exactly correspond to any of these, but seems to me to be the most attractive, is to set the baseline values at zero (by subtracting the utility of the baseline for each person from his other utilities) and then shrink the maxima to unity, shrinking the other utilities proportionally. 'Equal advantage' then simply means equal scores in terms of these (normalized) utilities. When, subsequently, I refer to the 'equal advantage' outcome I shall mean the one produced by this method of transforming utilities.

The Nash solution is rather different. It does not require any transformation of the utilities. In its simplest form it requires that we find the point on the Pareto frontier that maximizes the product of the utilities after we have subtracted from each utility the baseline score. Graphically, we get there by constructing a pair of axes with the baseline as their origin and then drawing horizontal and vertical straight lines off these axes to meet the Pareto frontier at a common point, such that the area enclosed by the resultant rectangle is at a maximum. The point on the Pareto frontier forming a corner is the Nash equilibrium. For the case where the Pareto frontier is a straight line cutting the two axes, the largest enclosed rectangle will always be achieved when the corner of the rectangle is half-way along the Pareto frontier. In this case, therefore, the Nash solution is equivalent to an equal-advantage solution obtained by making the axes equal in length and drawing a 45° line from the baseline. Where the Nash solution differs is in prescribing that the *same* point should be picked out even if only part of the Pareto frontier is a straight line, provided that the part that is a straight line contains the midpoint of a line extrapolated to cut the axes. (See Bartos, 1967, pp. 253–5.) This illustrates, incidentally, why the Nash solution uniquely satisfies the 'independence of irrelevant alternatives' condition. However, the general case of the Nash solution can still be regarded as a technique for preserving equal advantage over the baseline if we follow Harsanyi's idea that the Nash equilibrium is the point that (on certain assumptions about rational psychology) equalizes the willingness of the parties to risk non-agreement if their demands are not met. (Harsanyi, 1977, pp. 149–66; for a clear graphical development of the 'equal concession' justification of the Nash solution, see Nicholson, 1970, pp. 74–5.)

What are the actual answers to the problem of Luke and Matthew generated by these various solutions? Not surprisingly, they differ. But they share a common feature: all of them have Matthew playing over half the time.[5] Specifically, Raiffa has Matthew playing about 70% of the time, Braithwaite has him playing about 60% of the time, and Gauthier has him playing about 55% of the time. (See Luce and Raiffa, 1967, p. 149 and Gauthier, 1974b, p. 64.) The Nash solution, with any baseline, also gives us Matthew playing over half the time (87% of the time with cacophony as baseline, 60% of the time with the joint maximin as baseline, and 70% of the time with silence as the baseline.)

According to Braithwaite (1955, p. 37). "Matthew's advantage arises purely from the fact that Matthew, the trumpeter, prefers both of them playing at once to neither of them playing, whereas Luke, the pianist, prefers silence to cacophony." It is an interesting illustration of the way in which an oversimplification can be carried over from one writer to another that this statement is repeated uncritically by Luce and Raiffa (1957, p. 149), Sen (1970, p. 122) and Rawls (1971, p. 134), as if it applied to all solutions of the problem.

In fact, Matthew's advantage in playing-time also follows from the fact that Luke is more tolerant of Matthew's playing solo (relatively to playing solo himself) than is Matthew of Luke's playing solo. (Luke's play/listen ratio is 7:4 whereas Matthew's is 10:3.) This second condition is sufficient by itself to produce an unequal distribution of time. We could even go further: leave Matthew with the 'threat advantage', take cacophony as the baseline (so that he can cash in his 'threat advantage') and still finish up with Luke playing over half the time because he has to play a lot in order to gain 'equal advantage'. This would require that Luke should put listening to Matthew low while Matthew rates listening to Luke almost as high as playing himself. Thus, suppose we have a payoff matrix as in Table II, which satisfies these conditions. Taking cacophony as the baseline, we find that the Nash solution to this has *Luke* playing all the time, and the equal advantage criterion provides for him to play 80% of the time. Thus, Matthew's 'threat advantage' is swamped by what we might call Luke's 'whine advantage'.

TABLE II

		Matthew	
		Play	*Not play*
Luke	*Play*	(1, 2)	(11, 10)
	Not play	(3, 12)	(2, 1)

Of those who have criticized Braithwaite's proposed solution as unfair because it reflects Matthew's 'threat advantage', only Lucas genuinely stays on the target. As we have seen, he claims that the arbitrator should not take account of the damage each can gratuitously inflict on the other but should set the baseline at the best each can do for himself acting independently, allowing an equal gain to each above that level.

> Thus my external morality. It is hard, as are all legalistic moralities in which people do not care for one another and extend to one another only that consideration which is due to all men merely for being human, but it is less harsh than Professor Braithwaite's. There is no vindictiveness, only indifference. (Lucas, 1959, p. 10.)

Lucas's proposed solution – equal advantage in relation to the baseline – is identical in essentials to that of Gauthier, differing only in that Lucas takes over Braithwaite's transformation of utilities while Gauthier has his own technique. They agree in stipulating that Matthew and Luke should divide the time so as to retain the same relative advantage that they would have at the joint maximin point.

Gauthier's scheme is designed explicitly, he says, to be appropriate to 'civil society', the kind of society whose characteristic institution of social cooperation is the market (Gauthier, 1974b), and the psychology that Lucas depicts, of a passionless maximizer of utility (whose utility depends only on what happens to him), fits in well with this kind of society. That they come out in the same place is therefore to be expected.

I have argued above that the joint maximin may not be the appropriate baseline and that one might instead interpret the 'status quo' as one where neither plays. The important point for the present, however, is that both accept that it is fair for the division of time to be determined by taking the division that preserves equal advantage over an appropriate baseline. Gauthier adds that if we don't like this conclusion we have to rethink our attachment to civil society. (Gauthier, 1975, pp. 430–3; 1974b, pp. 62–3; and see also 1977, pp. 159–64.) And Lucas, while apparently feeling no qualms about his 'external morality' for most dealings with his fellow men, does say that there are alternative moralities appropriate to relations between monks or between lovers. (Lucas, 1959, pp. 10–11.) But the point remains that both accept, as the logical solution for maximizing individuals, a solution that is a modified version of Braithwaite's own. It is this conclusion that is challenged by Sen and Rawls.

Both of them, like Lucas, say that what is objectionable in Braithwaite's solution is Matthew's being able to use his 'threat advantage' but, unlike Lucas, they employ the term in a sloppy way that conceals rather than clarifies their real objections. Sen notes (correctly) that 'the collective solution' is 'crucially dependent on the status quo point' and he says that, even if the Nash solution predicts adequately the outcome of bargaining,

this does not mean that the Nash solution is an ethically attractive outcome and that we should recommend a collective choice mechanism that incorporates it. (Sen, 1970, p. 120.)

Rawls says that

to each according to his threat advantage is not a conception of justice. It fails to establish an ordering in the required sense, an ordering based on certain relevant aspects of persons and their situation which are independent from their social position, or their capacity to intimidate or coerce. (Rawls, 1971, p. 134.)

Both confuse the issue by writing as if their primary objection is to threats (i.e. gratuitous nastiness designed to improve one's bargaining position) when it is in fact to any aspect of relative bargaining strength entering into the outcome. Thus, Rawls speaks of someone 'threatening' to hold out unless the other parties agree with his demand for a favorable outcome. (Rawls, 1971, p. 140.) But it seems to me that the normal objections to threats hardly apply to someone's simply stating the price below which he refuses to sell, for example.

Beyond this, however, Sen and Rawls go in fundamentally different directions. Sen's objection is to *any* baseline-dependent solution. He falls into the tradition of utilitarian thought, extended to allow for distributive considerations, as by modern consequentialists such as Smart (1973). What ethics is about, from this point of view, is getting the distribution of utility right, and a right distribution of utility is to be characterized in terms of features such as the total (or average) utility derived from the arrangement, the minimum anyone gets, the degree of equality between the utilities of different people, and so on. Obviously, there is no room here for any reference to baselines.

For Sen, an ethical judgement is one that satisfies the criterion of impartiality, defined operationally to mean that it prescribes an outcome that Matthew and Luke would both choose if they knew everything about the situation except which one was which. Thus, he says that

Matthew himself might concede that if he did not know whether he was going to be Luke or Matthew before deciding on a system of distribution of time he might well have ignored the threat advantage and recommended a more equal sharing of time. (Sen, 1970, p. 122.)

Once we get to here, we have a straight run down a fairly familiar track. The premises, although they may appear minimal, in fact impose such strong constraints on the solutions possible that all of them bear a strong family resemblance, and must do because they can only take the form of specifying an abstract principle for the aggregation or distribution of utility. Thus, Harsanyi's theory of 'ethical judgements', in which an ethical judgement is one a person would make if he had an equal probability of being in any position, leads to utilitarianism (i.e. maximizing the average of the positions, weighted for the number of people in each). (Harsanyi, 1955, 1975; 1977, pp. 42–83.) Hare's principle of universalizability, when suitably generalized so that 'what I wouldn't want others to do' is taken as 'frustrate my wants', is said by its author to yield utilitarianism with some rather vaguely expressed egalitarian side-constraints (Hare, 1963). Once we get into the swing of things, Rawls's difference principle (expressed in terms of primary goods) is easily translated into a principle for the distribution of utility. Thus, Sen characteristically says that "on Rawls's analysis it turns out that the proper maximand is the *welfare* of the worst-off individual". (Sen, 1970, p. 136, my italics.) From the present perspective, these disputes are family quarrels.

On the assumption that the numbers of Braithwaite's original payoff matrix represent cardinal interpersonally comparable utilities, utilitarianism (maximizing the average) requires that Matthew play every evening because that maximizes the average (14 units to be shared between Matthew and Luke). Maximin and equality are identical in their requirements because, as we move along the Pareto-optimal frontier from Matthew playing all the time to Luke playing all the time, Matthew loses utility and Luke gains it, so the highest minimum occurs where the curve (actually a straight line) of Matthew's declining utility cuts that of Luke's increasing utility. The solution comes where Luke plays three-fifths of the time. This gives each of them an average of 5.8 units over each five-night sequence. (Matthew gets an average of 4 from playing and 1.8 from listening; Luke gets an average of 4.2 from playing and 1.6 from listening.)

It is always a sign of trouble when people's official theories fail to correspond to what they unreflectingly say about them. We saw that Sen said that Matthew might concede a 'more equal' division of playing time if he didn't know which of the men he was. Yet on the Harsanyi theory (which Sen explicitly contrasts with the bargaining approach), Matthew would demand to play *all* the time if he didn't know which he was.[6] And if Matthew and Luke agreed to go for equality of welfare or maximin welfare, they would finish up with a 3-2 split, just as proposed by Braithwaite, except that it would go in the other direction.

Rawls differs from Sen in two respects. First, although he agrees that bargaining strength should not enter in, he does not reject the relevance of a baseline. In this respect he is aligned with the 'fair division' theorists against Sen. Rather, what Rawls says "is lacking [in Braithwaite] is a suitable definition of a status quo that is acceptable from a moral point of view" (Rawls, 1971, p. 134, n. 10). The appropriate status quo is one that people would choose from behind a veil of ignorance, which concealed from them their particular preferences. I believe that it is in accordance with Rawls's analysis to suggest that the baseline that would be chosen would be one in which nobody was inflicting harm on anybody. In the case of Luke and Matthew this would entail that the baseline should be silence.

The second way in which Rawls differs from Sen is one that puts Sen on the same side of the fence as the 'fair division' theorists and sets Rawls apart from all of them. As we saw, Sen accepts the premise that, although the solution to Matthew and Luke's problem must be expressed in terms of a division of time between them, what we are really concerned with is a fair division of utility (want-satisfaction, 'welfare', happiness, etc.) between them. The stipulated distribution of time is one designed to generate the fair distribution of utility.

In this, however, Sen is parting from Rawls, for whom the primary goods (liberties, guaranteed opportunities, income etc.) are not to be treated as surrogates for the want-satisfaction to which they may or may not give rise. Principles of justice for Rawls *are* principles for the correct distribution of primary goods; it is not that principles of justice are really principles for the distribution of want-satisfaction and that it is merely for convenience that we express them in terms of primary goods. It is a mistake to take Rawls as saying here that we must have publicly-applicable principles and that degrees of want-satisfaction are too subjective to serve as the official distribuend. Admittedly, he sometimes says things that suggest this as for example where he speaks of primary goods as the 'clearest' basis of comparison (1971, p. 174). But his real point is that it is a ground-floor meta-principle of justice (criterion for something's counting as a principle of justice) that it should not go behind rights, opportunities, income etc., to take any note of want-satisfaction. (See especially Rawls, 1975, pp. 551–4.)

Rawls has not put forward a specific answer to the question of how Luke and Matthew should divide the evenings between them. It is, however, quite possible to develop an answer along Rawlsian lines. Whether or not Rawls would assent to the way in which I develop this answer, it is, I wish to claim, the correct answer to the question: What would be a fair division of the time?

Rawls, as I have mentioned, expresses his basic ideas in terms of a hypothetical choice from behind a veil of ignorance. The features of the veil of ignorance that Rawls introduces – the exact limitations on what things the parties are allowed to know about themselves and about the world – are justified by Rawls as ways of representing the 'constraints of right', that is to say the requirements that any just principles should satisfy.

Although I do not think anyone (including myself) has got to the bottom of it yet, my present inclination is to think that this was a mistaken strategy on Rawls's part. It has diverted, and continues to divert, effort and attention from the serious question – whether Rawls has got the constraints right and what principles of justice follow from them – and focuses it on the essentially unproductive question of whether the constraints can be expressed by adding to or subtracting from the veil of ignorance, and whether anything definite in the way of principles can be got out deductively from the notion of rational choice from the 'favored interpretation' of the original position. I shall therefore skip directly to the 'constraints of right'. My idea is that much of what Rawls tries to say by the clumsy device of talking about the construction of the original position can be better understood by talking directly about the constraints.

I shall begin by positing three 'constraints of right' as constitutive of any fair division. The first is that the baseline against which a fair division is to be measured is one in which neither Matthew nor Luke is harming the other. 'Harm' is defined here (as I proposed earlier) in such a way that it can be established without any knowledge of the actual preferences of Matthew and Luke. In the present case, this entails that the baseline be taken as being constituted by neither playing, because some people object to the sound of a musical instrument being played in an adjacent room that is not sound-proofed. Clearly, a lot more needs to be said, for a full account of the principle, about the characterization of harm. Not any dislikes are relevant: the dislike must be non-idiosyncratic and substantial to provide a basis for a legitimate interest in preventing it. Luke's possible mental distress at Matthew's sex life (provided, of course, it is not 'louder than a conversation') is not relevant, for example. For present purposes, however, it is not necessary to elaborate further, because I think that the playing of a musical instrument under the circumstances of Matthew and Luke falls in the core of the principle's application rather than the penumbra. The principle itself is, of course, a quite standard liberal one and I have nothing to add to the standard arguments in its favor.

It should be emphasized that calling silence a baseline is a normative proposition. It does not depend on the actual status quo existing at the time that a determination of a fair division is made. The point is worth making here because Sen, for example, moves from observing that the status quo may be ethically objectionable to looking for baseline-independent solutions (or 'social welfare functions'). The tendency to use the notion of the status quo both for the actual state of affairs that obtains and for an ethically acceptable baseline is endemic, and makes confusion inevitable.

The second 'constraint of right' is that the object of distribution is understood to be time and not utility. As I have explained, this does not merely mean that the solution should be expressed as a division of time. (This was true of all the other solutions, too.) It means that the rationale of the solution is not to depend on any reference to the distribution of utilities that it is designed or expected to produce. That is to say, the solution cannot be specified in such a way that it involves comparisons between the absolute utilities of Matthew and Luke, e.g. 'Luke gains more than Matthew loses' or 'Luke is as well off as Matthew'. Nor can it involve any references to the relative utilities of Matthew and Luke, e.g. 'Luke more strongly prefers X to Y over Z than Matthew does Y to X over Z' or 'Luke prefers Y to a fifty-fifty chance of X or Z, whereas Matthew prefers the lottery to Y'. The only information about utilities that can be used is simple preference-orderings, e.g. 'Luke prefers X to Y'. Thus, suppose we are to divide a cake which is to be eaten on the spot. The thesis is that we can divide it fairly without knowing how much the different claimants like different amounts of cake, but we must know that one person prefers no cake to any, another does not want more than a slice of a certain size, and so on.

I should emphasize that this constraint operates only if it is understood that Matthew and Luke are arguing about what is required for a fair division. It would be generous, or beneficent, for one to yield more to the other than this, and in some circumstances downright curmudgeonly not to, but it would not be unfair to refuse to. Or to look at it from the other side, fairness establishes the maximum that each can claim from the other as of right, though that leaves it open to either to make an appeal on an alternative basis — friendship, sympathy, services rendered in the past, and so on.

The best evidence I can offer is perhaps that, in spite of their official theories, those who have discussed the other solutions have in fact gone straight to the division of time rather than the division of utility, and put their intuitions to work on that. More generally, nobody who has escaped long exposure to modern economics or modern philosophy would suppose for a moment that the fair distribution of money is anything but a fair distribution of the chance to buy things or that the distribution of playing time is anything but the distribution of time in which to play. The fact that even those who officially abandon this view cannot really shake it off is strong evidence in favor of its intuitive appeal.

The third 'constraint of right' is that, in the absence of any special reason to the contrary (and the second constraint entails that nothing involving the utilities in the payoff matrix can count as a special reason) a fair division is an equal division, relatively to the appropriate baseline. This can be supported abstractly: it is indeed something close to a tautology. It can be supported by an appeal to usage: 'fair' and 'equal' can often be interchanged in distributive contexts. And out of the mouths of babes and sucklings we hear the cry 'It's not fair' go up whenever access to some toy is not equally shared.

Putting the three constraints together, we get, I believe, the following specification of a solution to the problem of Matthew and Luke:

> *A fair outcome is one that divides the time between solo playing and silence in such a way as to satisfy the following two conditions: (a) any time used for solo playing is divided equally between Matthew and Luke; and (b) both Matthew and Luke prefer the proportion of time earmarked for solo playing to the outcome constituted by neither ever playing.*

It should be noted that this solution applies only to cases where both Matthew and Luke have playing solo as their first preference, so that the problem is indeed one of dividing up the right to play solo. If one of the players prefers silence to anything else, there is no problem of fair division because the first constraint of right immediately gives the player silence, which is what he most wants, and there is nothing to discuss (at any rate in terms of fairness). If one most prefers playing solo and the other most prefers listening, there is no conflict of interest, so again no problem of fair division. (See Axelrod, 1970, p. 5.) However, the solution does not restrict itself to cases in which both players put listening to the other second after playing solo, as we shall see.

I believe that this formula for a solution uniquely satisfies the three 'constraints of right' that were postulated. The requirement of equal playing time [condition (a)] derives from the baseline's being silence (first constraint) and the fact that 'equal advantage' relatively to it (third constraint) must be defined in terms of playing time rather than utility (second constraint). That the outcome, if it involves any playing time, must be preferred by both to silence [condition (b)] follows immediately from the concept of a baseline (first constraint).

I have offered some general arguments for the three 'constraints of right'.

The other way to defend them is, of course, to show what implications they have and to argue that these implications seem reasonable. Assuming, therefore, that the solution I have deduced from them is correctly deduced, the next step is to apply it to the story of Matthew and Luke. I shall first discuss the story as told by Braithwaite, and then two simple variants on it, generated by changing certain cells in the payoff matrix.

One of the artificialities of the solutions considered so far is that they set no limits on the fineness with which the proportions of evenings devoted to different things could be divided up. Let us instead suppose that Matthew and Luke agree on Sunday as a day of rest, and also agree that whatever proportions of silence and solo playing are settled on should be contained within a week. This means that there are only four possible outcomes consistent with the solution: Matthew and Luke play three evenings a week each, two each, one each, or not at all. We shall need to ask what the preferences of Matthew and Luke between these alternatives would be. We can use the utilities in the relevant payoff matrix to answer this because all that will enter into the solution is a preference-ordering for each. We do not ever compare Matthew's and Luke's relative payoffs, even in the way required by the equal advantage or Nash solutions.

First Example: Braithwaite's Utilities

It is a consequence of the preference-orderings postulated by Braithwaite between playing, listening and silence (or cacophony) that both Matthew and Luke will have the same preference-order between six nights of solo playing equally divided, four nights, two nights, and none. (The payoffs are shown graphically on Figure 1, which also shows the Pareto-optimal points and connects them to show the Pareto-optimal frontier.) Thus, if we add a requirement of weak collective rationality, specifying that no point will be chosen if there is another that both prefer to it, there is a unique outcome produced by our two criteria of fairness, namely that corresponding to Luke and Matthew each playing three nights a week. (It is circled in Figure 1.)

The condition of weak collective rationality that I have added to the original solution is, I believe, non-controversial. I have, indeed, already assumed it a little earlier by saying that if Luke and Matthew both coincide in their highest preferences — that is to say, if the same thing suits them both best — no problem of fairness arises because there is no basis for disagreement.

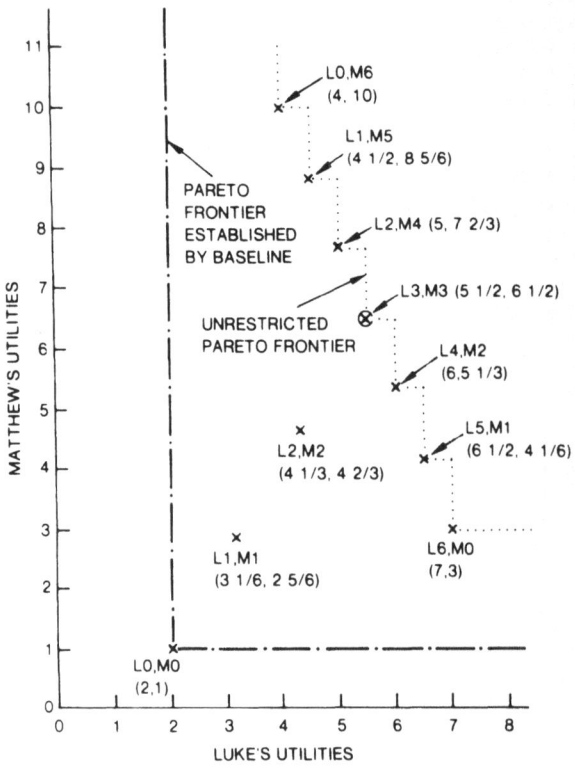

Fig. 1.

The imposition of an equal distribution of playing time (leaving open only the question of the division between playing time and silence) eliminates the conflict of interest that is otherwise inherent in the situation with Braithwaite's payoffs. Out of the alternatives now available, both agree on the same one as the most preferable.

I would suggest that in the case as set out by Braithwaite, and any other in which Matthew and Luke prefer both playing and listening to silence, the answer that they should split all the available time equally is obviously right. (Non-systematic survey evidence that I have collected by simply asking friends what they think strongly confirms this.) The solution I have proposed will always produce the answer (for any payoff matrix in which the order of preference among outcomes is preserved) that Matthew and Luke should play half the evenings each. No other solution that I have discussed can claim this, and in my view no solution that fails in such a simple case deserves to be taken seriously. Even if my solution is not finally acceptable, I think it is clear that the others can be rejected out of hand. It is striking evidence of the way in which a certain mind-set can be created by the way in which a problem is looked at, that nobody who has discussed the problem before has started from the proposition that any solution not entailing that Matthew and Luke play half the time each clearly fails to articulate our basic intuitions about fair divisions.

Second Example: Luke Prefers Silence to Listening

It might be thought that my solution is unnecessarily complicated. Why not simply prescribe that, whatever the payoffs may be, Matthew and Luke play on half the evenings each? To illustrate the inadequacy of such a simple solution, consider the following modification of Braithwaite's story.

Suppose now that Luke, instead of getting 4 units of utility from listening to Matthew, gets -4. He may perhaps dislike cacophony even more, but his utility from cacophony plays no part in the discussion so we can disregard it. The important thing is that he now prefers silence to listening to Matthew play. Indeed, he prefers two nights of silence (which still yield him 2 units of utility per night) to one of playing and one of listening (which yield an average of one and a half units). If Matthew's utilities are as before we get Figure 2. Since Luke can insist on silence if he wants it, and he does prefer it to any equal division of playing time, the outcome produced by my solution is that neither plays on any evening.

It may be seen, however, that this outcome is not Pareto-optimal in that there are ways of dividing up the playing time *unequally* that both Matthew and Luke prefer to silence every evening. In Figure 2, the Pareto frontier contains the outcomes in which Luke plays four times out of six, five times out of six, and six times out of six.

The question we have to ask now is: if we assume (as we have done) that a weak conception of collective rationality entailed choosing three nights of playing each rather than two nights each in the first example, doesn't it entail that Matthew and Luke should choose a point on the Pareto-optimal frontier if both prefer it to the outcome prescribed by the requirement of equality? If we take strategy into account the answer should be 'Not necessarily'. But if

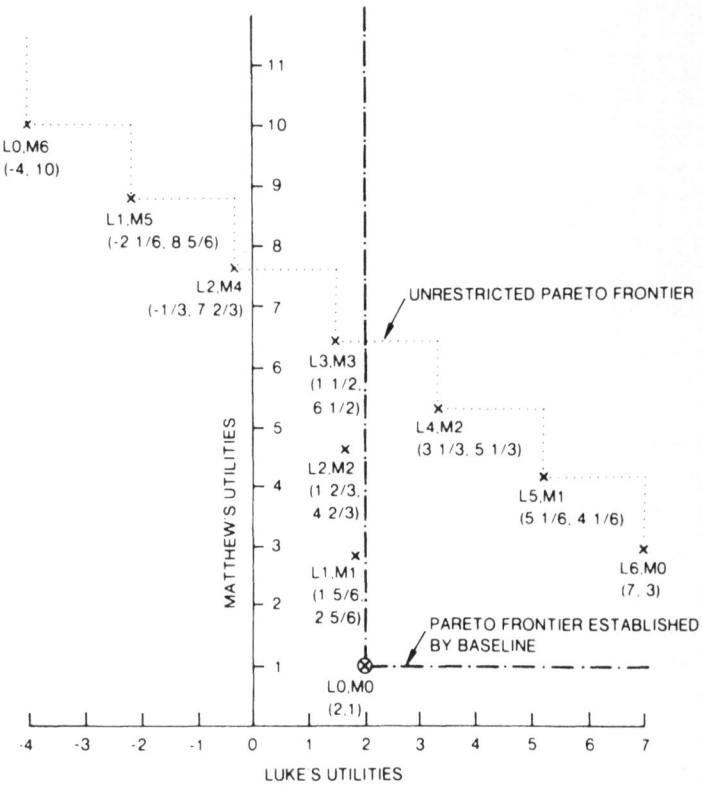

Fig. 2.

we ignore strategic considerations it does look like a piece of collective irrationality for Matthew and Luke never to play when there are several possibilities that both would prefer.

I therefore propose to call the solution put forward earlier the *strict solution* and talk of an *extended solution*. The extended solution is as follows: we first establish the outcome(s) prescribed by the strict solution, that is to say the undominated equal outcomes. Then we see if any such outcome is within the Pareto frontier. If it is, then we replace that outcome with the outcome or outcomes that are on the Pareto frontier and preferred by both Matthew and Luke to it. In the present case, as we have seen, there are three points on the Pareto frontier and any of them is, in the extended sense, a fair outcome.

If there is a danger that the parties, by the obstinacy of their bargaining, may fail to reach a Pareto-optimal point, they could call in an adjudicator to pick a point on the Pareto-optimal frontier for them. In such a case, there is no reason why he should not use a predictive model to come up with a solution, since the object can plausibly be represented as being to pick that fair outcome that the parties would have agreed on by bargaining rationally.

Third Example: Luke Prefers Silence to Listening and Has Diminishing Marginal Utility from Playing Solo

We might expect that either Matthew or Luke would, with some payoff structures, most prefer some playing and some silence (among the equal-time outcomes) to either all playing or all silence. However, Braithwaite's stipulation that their appetite for playing on any given evening is not affected by what has happened on previous evenings has the effect of ruling out such a possibility. Either the average payoff of playing and listening for equal periods exceeds that of silence, in which case the actor always prefers more playing time to less or silence is superior to a fifty-fifty mix of playing and listening, in which case the less playing the better. If the average of playing and listening is the same as silence, any proportion of playing time is equally good.

To get in a possibility that someone might prefer an intermediate amount of playing within the constraint of equality we have to allow his utility of playing to vary inversely with his frequency of playing. (Or we could make the value of silence increase the less of it is experienced, which is in effect to say that both playing *and* listening both pall with repetition.) Suppose, for example, that the value to Luke of an evening of playing varies with the number of evenings in a week that he plays as follows: the first night yields 12 units, the second 9 units, and each extra night 3 units less. His utility of listening to Matthew is -4 units and silence is, as usual, worth 2 units. This results in Luke most wanting (among the equal-time possibilities) two nights each of playing and two of silence, then one night each of playing and four of silence, then three nights each of playing and no evenings of silence, and last silence all the time. If Matthew has the same utilities as usual, we have Figure 3. We can see that silence all the time comes in last place for both, and also that playing one night each is dominated by playing two nights each. We can therefore eliminate these as outcomes under the strict solution. But playing two nights each does not dominate playing three nights

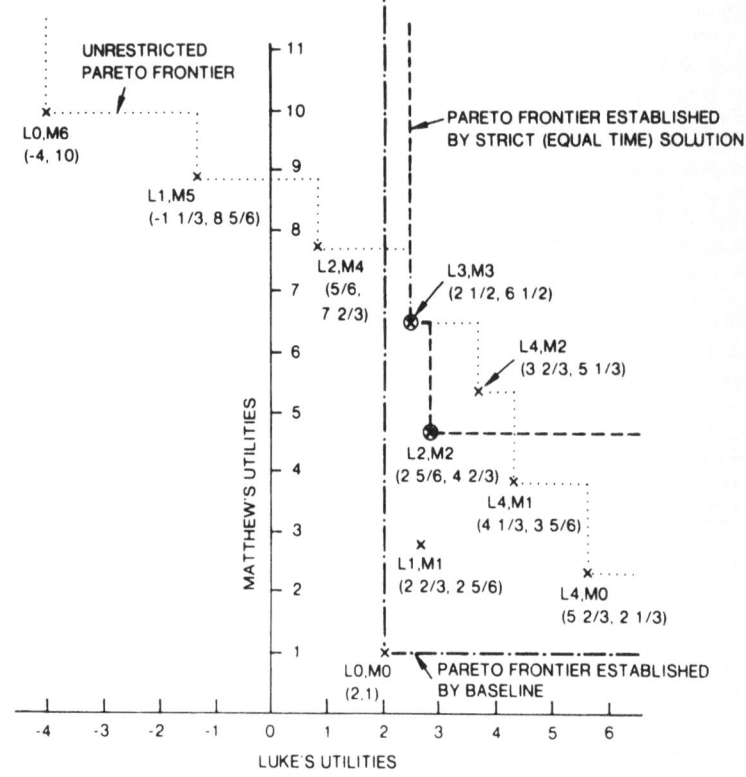

Fig. 3.

a week, nor is it dominated by playing three nights a week. Luke prefers the
first and Matthew the second. Both outcomes therefore stay in. They are
equal-time outcomes; they are outcomes preferred to the baseline by both
Matthew and Luke; and neither outcome is dominated by the other. It seems
to me quite reasonable here to say that whichever outcome they reach by
bargaining (or by an adjudication intended to simulate the outcome of bargaining) is fair as well as collectively rational.

We can ask whether the extended solution has any application here. Since
we have, in the present case, two alternative outcomes produced by the strict
solution, the question becomes whether *either* of them is dominated by some
other outcome. As we can see from Figure 3, the outcome preferred by
Matthew, where both play three nights a week, is already on the Pareto-
optimal frontier. But the outcome preferred by Luke, where they both play

two nights a week, is inside the Pareto-optimal frontier. It is dominated by the outcome where Luke plays four times a week and Matthew two. A moment's reflection tells us that this must be a mutual improvement because Luke prefers playing to silence and Matthew prefers listening to silence. The extended solution thus turns out to produce two outcomes one of which is the same as in the strict solution. As between those two it is again a question of bargaining or adjudication to simulate the results of bargaining. Either outcome is, in our extended sense, a fair outcome.

The three examples illustrate the range of difference it can make to move from the strict to the extended solution. In the first example, the solution prescribes a point on the Pareto frontier, so allowing for the extended solution made no difference. In the second case, the solution prescribes silence, and the extended solution permits any outcome on the Pareto frontier that is better for both than silence. The extended principle thus operates in exactly the same way as a principle simply stating that any outcome preferred by both parties to the baseline is to be reckoned as fair. The third example lies between the other two in the scope that it gives the extended solution. Although the extended solution permits a move to a point on the unrestricted Pareto frontier that is not an equal-time position, there are other points on the Pareto frontier and better for both than the baseline that are ruled out. Note the points L4, M1 and L4, M0 in Figure 3: these are ruled out because they are both worse for Matthew than the less favorable of the two equal-time outcomes prescribed by the strict solution, (L2, M2). Thus, permitting the extended solution to introduce a new outcome by no means has to entail that the initial fair solution will fail to constrain the outcome tightly.

At this point in the paper, space considerations prevent me from succumbing to the temptation to gallop off in all directions. Obviously, nobody really cares about Luke and Matthew and their ridiculously artificial predicament. The problem they pose is worth discussing only because it offers the prospect of making some progress with the real problem of distributive justice in groups, in societies and in the world as a whole. On this I shall offer some brief and I hope tantalizing remarks.

The generalization of the extended solution to n persons gives us the difference principle in the form: 'Everyone must gain from an inequality'. But it does not give us the specific form espoused by Rawls in *A Theory of*

Justice, that the worst-off must be as well-off as possible. (See Rawls, 1969, pp. 63–9 for a discussion of alternative interpretations of the general formula.) Thus, applied to my second version of the Luke and Matthew story, the general difference principle – that everyone should gain from an inequality – permits any of the outcomes that are mutually advantageous in comparison with silence: Luke playing four times, five times or six times a week and Matthew on the remaining evenings. But the specific (maximin) form of the difference principle would prescribe that the first of these is alone fair because it maximizes the amount of primary goods (in this case playing time) obtained by Matthew, who gets less under all three arrangements.

The opportunity to play a musical instrument without interference from extraneous noise is, I think, a genuine case of a primary good in Rawls's terms. Like any other opportunity it is in general rational to prefer more of it to less, since even if you do not find that you actually want to take advantage of it you are no worse off than you would be without the opportunity, and if you do want to then you are glad to have it. The case of Luke and Matthew is, of course, one in which Luke's enjoyment of the primary good of playing solo imposes a cost (at any rate an opportunity cost) on Matthew and vice versa. But that is in fact rather characteristic of primary goods. The strict principle of fairness says, in effect, that a primary good should not be allowed to anyone if an equal distribution of any positive amount of it imposes harm on anyone. The extended principle than allows those who are harmed to waive their veto if they are adequately compensated. To avoid extortion and bargaining costs, a public authority may have to decide what is adequate compensation, but the principle that harm must be compensated for remains.

I believe that the 'constraints of right' generate only the weak, general form of the difference principle, and that the strong form of the difference principle comes out of an incompatible theoretical approach, that of constrained maximization under conditions of uncertainty. The device of the original position is employed to yoke together the 'constraints of right' and constrained maximization, but does not succeed. Depending on what aspect of Rawls's theory one is more impressed with, it is possible to come out with sharply differing reconstructions of it. I suggest that this helps to account for the richness and difficulty of *A Theory of Justice*. These remarks are hardly conclusive (even if they conclude) but I hope they serve to indicate some of the larger issues opened up by the fable of the pianist and the trumpeter.

University of Chicago

NOTES

[1] Acknowledgement of financial support for the preparation of this paper is gratefully made to the Raymond W. and Martha Hilpert Gruner Social Sciences Endowment in the University of Chicago. This paper has benefited from many comments, but I am particularly indebted to Howard Sobel for a suggestion that I have incorporated.

[2] See Harsanyi (1977), pp. 167–79 for a discussion of optimal threat strategies within the Nash framework which concludes (p. 179) that the best strategy is one that represents "the best possible compromise between trying to *maximize* the *damage* that one can cause to the opponent in a conflict situation and trying to *minimize* the *cost* of the conflict to oneself". This analysis is limited by the assumption that threats have to be carried out if their terms are not met and also the usual Nash assumption that "the players have full information about each other's utility functions". (Harsanyi, 1977, p. 190.) But the general conclusions stand, I believe, in the absence of these limitations.

[3] Lucas is not explicit that what he has in mind is the joint maximin but speaks of the appropriate baseline as the one where Matthew and Luke are both pursuing their 'prudential strategies', understanding a prudential strategy for each as "the one best designed to secure his own interests" (Lucas, 1959, p. 9). However, we can deduce that it is intended to be the maximin because he describes his preferred baseline as the point of "the intersection of the horizontal and vertical tangents of the parabola" and this is in fact the point corresponding to the joint security level of Matthew and Luke. (Lucas, 1959, p. 9.) See Luce and Raiffa, 1957, Fig. 15, p. 149, where all the relevant lines are drawn in.

[4] The way this works out is as follows. With Luke playing one (random) night in four and Matthew one (random) night in five, Matthew gets, on the average, over a period of 20 evening of cacophony (2 units), twelve evenings of silence (1 unit), three evenings of solo playing (10 units) and four evenings of listening to Luke play solo (3 units). This averages to 2.8 units, which is Matthew's maximin value.

[5] Lucas is an exception, claiming that an equal-advantage move from the joint maximin would have Luke playing more often. However, this entails, as Lucas says, "accepting the slope of the isorrhopes [45° lines] as giving the measure of equal increments of utility" (Lucas, 1959, p. 9), yet if Lucas is going to throw out the rest of Braithwaite's argument there seems no point in keeping his peculiar way of transforming the utilities. A 45° line from the joint maximin to the Pareto-optimal frontier leaves Matthew playing a majority of the time on any of the other methods of transforming the utilities that I have mentioned.

[6] It may be intuitively repugnant to conclude that average utility is maximized if Luke never plays, but it should be observed that it is specified in Braithwaite's statement of the case that the utility of playing is linear with frequency of playing, in other words there is no diminishing marginal utility of playing. In von Neumann and Morgenstern terms, this means that each is indifferent between the certainty of playing one evening and a $1/n$ chance of playing n evenings, where n is a whole number. It may be noted that this is the kind of utility function out of which the St. Petersburg paradox is generated.

BIBLIOGRAPHY

Axelrod, R.: 1970, Conflict of Interest (Markham, Chicago).

Bartos, O. J.: 1967, Simple Models of Group Behavior (Columbia University Press, New York).

Braithwaite, R. B.: 1955, Theory of Games as a Tool for the Moral Philosopher (Cambridge University Press, London).

Gauthier, D.: 1974a, 'Justice and natural endowment', Social Theory and Practice 3, 3–26.

Gauthier, D.: 1974b, 'Rational cooperation', Noûs 8, pp. 53–65.

Gauthier, D.: 1975, 'Reason and maximization', Canadian Journal of Philosophy 4, pp. 411–33.

Gauthier, D.: 1977, 'The social contract as ideology', Philosophy and Public Affairs 6, pp. 130–64.

Hare, R. M.: 1963, Freedom and Reason (Clarendon Press, Oxford).

Harsanyi, J. C.: 1955, 'Cardinal welfare, individualistic ethics, and interpersonal comparisons of utility', Journal of Political Economy 63, pp. 309–21.

Harsanyi, J. C.: 1975, 'Can the maximin principle serve as a basis for morality? A critique of John Rawls's theory', American Political Science Review 69, pp. 594–606.

Harsanyi, J. C.: 1977, Rational Behavior and Bargaining Equilibrium in Games and Social Situations (Cambridge University Press, London).

Lucas, J. R.: 1959, 'Moralists and gamesmen', Philosophy 34, pp. 1–11.

Luce, R. D. and Raiffa, H.: 1957, Games and Decisions (Wiley, New York).

Nash, J. F.: 1950, 'The bargaining problem', Econometrica 18, pp. 155–62.

Nicholson, M.: 1970, Conflict Analysis (The English Universities Press, London).

Rawls, J.: 1958, 'Justice as fairness', Philosophical Review 67, pp. 164–94.

Rawls, J.: 1969, 'Distributive justice', in Peter Laslett and W. G. Runciman (eds.), Philosophy, Politics and Society, 3rd Series (Blackwell, Oxford).

Rawls, J.: 1971, A Theory of Justice (Harvard University Press, Cambridge, Mass.).

Rawls, J.: 1975, 'Fairness to goodness', The Philosophical Reveiw 84, pp. 536–54.

Sen, A. K.: 1970, Collective Choice and Social Welfare (Holden-Day, San Francisco).

Smart, J. C. C.: 1973, 'An outline of a system of utilitarian ethics', in J. C. C. Smart and Bernard Williams (eds.), Utilitarianism: For and Against (Cambridge University Press, London).

THOMAS SCHWARTZ

WELFARE JUDGMENTS AND FUTURE GENERATIONS

ABSTRACT. The author argues that long-range welfare policies – policies designed to provide significant, widespread, continuing benefits to future generations, remote as well as near, at some cost to ourselves – cannot be justified by appeal to the welfare of remote future generations. He questions whether they can be justified at all. The problem is that the failure to adopt such a policy would not make any of our distant descendants worse off that he would otherwise be, since had the policy been adopted, *he* would not even have *existed*. These considerations also bring out a conflict between utilitarian and Paretian principles.

We are continually urged to adopt any number of *long-range welfare policies*, as I shall call them – policies designed to provide significant, widespread, continuing benefits of some sort to future generations, remote as well as near, at, some cost to ourselves and possibly to earlier future generations.[1] Examples include policies of natural-resource conservation, environmental cleanliness, genetic health and variety, cultural preservation and population control.

I demur. I question whether long-range welfare policies can be justified at all. I argue mainly that none can be justified by appeal to the *welfare of remote future generations*.

1. THE CASE OF THE DISAPPEARING BENEFICIARIES

Let P be any long-range welfare policy. Applied to P, the position I shall attack goes something like this:

> Our distant descendants would likely be significantly worse off if P were not adopted than if P were adopted.

I will argue that this position is wrong if interpreted so as to be relevant to the justification of P.

The position can be interpreted two ways:

> WAY 1   At least some of our distant descendants would likely

Theory and Decision 11 (1979) 181–194.   0040–5833/79/0112–0181$01.40

be significantly worse off in some respect if P were not adopted than those *very same individuals* would be if P were adopted.

Objection: Suppose P is not adopted. Consider those of our distant descendants whose lives will have been significantly affected thereby. Let X be any one of them. Then X would *not have existed* had P been adopted and hence will be no worse off than he would have been had P been adopted.

For X to be one of those possible individuals, say Y, who would have existed under P, it is not enough that Y would have had X's name, else I would be identical to all the Thomas Schwartzes in the world. Nor is it sufficient that Y would have had X's ancestry, else my son and daughter would be identical. What *would* insure that Y would have been the same person as X – that X would have been Y, had P been adopted?

That is notoriously controversial. Happily, we have no need to enter the controversy. For it is quite certain that no one born under P would have been the same person as X, by *any* plausible criterion of personal identity.

P would have brought about a future world different from the (supposed) actual future world, and X is one of those significantly affected by P's non-adoption. So the circumstances of X's life would not have been fully replicated under P. One or both of X's parents might not have existed under P. Had they both existed, they might not have met. Had they met, they might not have mated. Had they mated, they might not have procreated. Even had they procreated, their offspring, Master Z, would have been conceived under conditions at least a little different from those of X's conception.

But that makes it virtually impossible for Z to have developed from the *same pair of gametes* as X, hence virtually impossible for Z to have had the *same genotype* as X. After all, trivial circumstantial differences could easily have determined whether intercourse, ejaculation or conception took place at any given time. And the most minute circumstantial differences – a difference of one degree in temperature, say, or of one drop of some chemical – would have determined *which particular spermatazoon fertilized the ovum* and *what particular pattern of meiosis was involved in the production of either gamete*. Let the circumstances surrounding Z's conception be as similar to those surrounding X's conception as one can realistically suppose, given my assumption about X and Z. Still, Z would have been no more likely than X's non-twin sibling to have had X's genotype.

The combination of a different genotype and a different environment insure all manner of further difference: Z would not have been composed of the same matter as X, he would not have looked the same as X, and he would not have had the same perceptions, the same memories, the same beliefs or attitudes, the same capacities, the same character or the same personality as X. As a result, Z would not have had the same psyche as X – by any sensible standard of sameness of psyche. Neither would Z have fulfilled just the same social roles and relationships as X.

In sum, Z would have differed from X in *origin* (different gametes), in *content* (different matter, different mind) and in *basic 'design'* (different genotype), and he would have been shaped by a different environment to perform different functions. We have no more reason to identify Z with X than we have to identify siblings reared apart with each other.

To put the point another way: Let the $+P$ *population* comprise those possible future people who would be born if P were adopted and significantly affected by P's adoption. Let the $-P$ *population* comprise those possible future people who would be born if P were not adopted and significantly affected by P's non-adoption. These two possible populations will quickly start to diverge and eventually diverge completely (I have, in effect, argued), so that no member of either will belong to the other.

Lest one think that the rate of divergence will taper off, precluding complete divergence, or that complete divergence will not occur within the remotely forseeable future, it is worth noting that once the two possible populations begin to diverge, the rate of divergence will increase exponentially.

To see why, let us divide the future, somewhat arbitrarily, into generations 1, 2, 3, etc. Let $p(i)$ be the probability that a randomly selected member of the $-P$ population born in the ith generation also would belong to the $+P$ population. Such a person won't belong to the $+P$ population unless both his parents belong. But the probability that his mother belongs is $p(i-1)$, and similarly for his father. Therefore, if membership by his mother in the $+P$ population and membership by his father in $+P$ population are completely independent events, then the probability that *both* his parents belong to the $+P$ population is $p(i-1)^2$, so that

(*) $p(i) \leqslant p(i-1)^2.$

One might contend that membership by our friend's mother in the $+P$ population and membership by his father in the $+P$ population are not completely independent events. But because meiosis, spermatazoon-selection and conception are determined in such subtle ways by such minute circumstantial factors, it is extremely unlikely that the birth of either parent under P would significantly increase the probability of the other parent's birth under P. And as I have argued, it is at best extremely unlikely that a member of the $-P$ population would belong to the $+P$ population *even if* both his parents belonged, circumstances being sufficiently different under P to preclude conception involving the very gametes whence he sprang. So (*) still is eminently plausible. If anything, (*) is too weak; if anything, $p(i)$ is *far less* than $p(i-1)^2$.

But according to (*), as soon as the two possible populations diverge a little, they will continue to diverge at an exponentially increasing rate: As soon as $p(i)$ is even a little less than 1, $p(i)$ will rapidly approach 0. For example, if $p(i) = 0.8$, (*) implies that $p(i + 6) \leqslant 0.0000005$. This means that even if, after several generations, there still are fully 8 chances in 10 that a member of the $-P$ population also belongs to the $+P$ population, the chances of this happening six generations later are at most 1 in 2 million. And if, as I have suggested, $p(i)$ really is a lot *less* than $p(i-1)^2$, then it shouldn't take even six additional generations for complete divergence to obtain.

> WAY 2   Although none of our distant descendants who would exist if P were not adopted also would exist if P were adopted, the society that comprises our distant descendants — the *distant future society*, for short — would itself exist whether or not P were adopted. One and the same society can have different individual members under different circumstances, actual and hypothetical. And while no *individual member* of the distant future society would be worse off if P were not adopted than *he* would he if P were adopted, the *distant future society itself* would be worse off. Its standard of living would be lower, its civilization inferior.

Objection: Let us concede that one and the same distant future society would exist whether P were adopted or not, although it had none of the same members under the two alternatives. Still, in the absence of P, this 'society' would not be worse off *in any ethically relevant sense*; it would not be worse off in any sense relevant to the justification of public policy. The fact that one policy alternative would in some sense be worse than another for something called a society, although in no sense worse for any *person*, constitutes no ethical ground for prescribing the latter alternative.

Some of those now alive may feel an urge to insure that the distant future society will be an appealing one in which to live, with an admirable civilization. And if sufficiently many people are sufficiently minded this way, that fact may constitute some sort of justification for some sort of long-range welfare policy. But this policy would be no favor to our *distant descendants*. *They* could not reproach us for having rejected it. The beneficiaries of such a policy would be *ourselves* – those of us, anyway, who get their kicks from the prospect of a flourishing future society.

In sum: *P* cannot be justified by appeal to the welfare of our distant descendants, because the failure to adopt *P* would hurt not a single one of them.

2. THE FALLACY OF POPULATION-CONFLATION

Long-range welfare policies owe their appeal to a special fallacy of ambiguity – the fallacy of Population-Conflation, to give it an appelation.

Consider the proposition that our distant descendants would be worse off if *P* were not adopted than if *P* were adopted. To contend that this justifies *P*, at least to some extent, is to contend that a certain proposition of the form:

(W)   *Ds* would be worse off if we did not do *A* than if we did *A*

provides some support for the corresponding proposition of this form:

(S)   We should do *A*.

Ambiguity lurks. (W) can be interpreted two ways:

(W1)   Those who are *in fact Ds* would be worse off if we did not do *A* than those *very same individuals* would be if we did *A*.

(W2)   Those who would be *Ds* if we did not do *A* would be worse off if we did not do *A* than those who would be *Ds* if we did *A* would be if we did *A*.

Surely only (W1) would, if plausible, provided any support for (S). But when *A* stands for the adoption of a long-range welfare policy, only (W2) has any plausibility (in view of §1). So when *A* stands for the adoption of a long-range welfare policy, the use of (W) to support (S) turns on an ambiguity: (W) would, if plausible, provide some support for (S), but under a different interpretation from that which renders (W) plausible. In general, when (W) owes its plausibility to (W1) rather than (W2), you cannot use (W) to support (S).

The same fallacy can be committed, by the way, in connection with policies that have little or nothing to do with future generations. Think of 'urban renewal' projects that are designed to make specified neighborhoods better off, but which do so, in effect, mainly by bringing in better-off residents, displacing many original inhabitants. Or consider Federal minimum wage laws. In a way, they raise the welfare level of the lowest-paid workers employed in 'interstate commerce', but they do so, in part, by altering the membership of that category, making many erstwhile members worse off, rather than by improving the lot of the original membership.

3. THE DECLINE AND FALL OF UTILITARIANISM

Utilitarianism, whether applied to individual actions, to social norms or to public policy, requires the maximization of social utility – construed as happiness, pleasure, welfare, or some such thing. If there is any case to be made for long-range welfare policies, it is most likely a utilitarian case. But problems beset both the utilitarian principle and its application to certain long-range welfare policies.

These problems appear when we try to choose on utilitarian grounds between two *population policies*: a severely *restrictive* policy, designed to limit population size, and a more or less *laissez-faire* policy, consisting of doing little or nothing about population size.

By insuring that there were fewer people to share limited resources, the restrictive policy might very well do more good on the average – insure people a larger per-capita share of utility – than would the laissez-faire policy. And one variety of utilitarianism enjoins us to do as much good as possible *per capita* – to maximize *average* utility.

But another, more traditional variety would have us maximize the *total* good we do – *aggregate* utility. And there is reason to believe that the laissez-faire policy would do more total good for people than the restrictive policy.

Under the laissez-faire policy, there would be less utility per person, perhaps, but many more people to share some measure of utility. There would be a larger pool of human resources (strong backs, skilled hands, clever minds) and greater economies of scale. There would be lower, more easily satisfied expectations and a much greater incentive to uncover hitherto hidden resources – including new uses for old resources, as well as territory, energy and raw materials found through under-sea, out-space and inner-molecular explorations. And because people would tend each to consume a smaller quantity of resources, those resources would tend to have a *greater marginal value*. Just as a dollar is worth more to a poor man than to a rich man, so a unit of resources would tend to be worth more to a resource-consumer under the laissez-faire policy than under the restrictive policy. As a result, while resources would (I suppose) yield less utility *per consumer* under the laissez-faire policy, they would likely yield more utility *per unit consumed*, and therewith more *total* utility.

I have heard it argued, though, that a laissez-faire population policy would lead to such an enormous growth of population that life would no longer be worth living. But if life really were not worth living – if people wished they had not been born – what would explain the existence of such a huge population? Even allowing for ignorance, lust and religious belief, how much incentive could there be to reproduce? It is hard to see how lives so wretched would replicate themselves to rapidly.

True, even the poorest people procreate – too much, by some lights. But who's to say their lives are not worth living? *You* may find their lives worthless. *You* may prefer no life to theirs. But for all their hardships, *they* don't seem to feel that way – not literally. In assessing aggregate utility, what is important is whether life is worth living to those who live it.

I do not say that the laissez-faire policy should be adopted, that utilitarianism requires its adoption, or that utilitarianism is properly interpreted in terms of total rather than average utility – as *aggregate* rather than *per-capita utilitarianism*, let us say. My only point is this: It is *quite possible* that aggregate utilitarianism supports the laissez-faire policy, even if per-capita utilitarianism supports the restrictive policy, hence quite possible that we would be unable to choose between the two policies until we'd chosen between the two versions of utilitarianism – or come up with a third version.

If the rationale behind utilitarianism is that we should do as much good as possible and that happiness (or pleasure or welfare or whatever) alone is good ('ultimately', 'intrinsically'), then it must be *total* happiness that the utilitarian would have us maximize.

I have heard utilitarian partisans of population control argue this way: What is good is not really happiness as such, but the happiness of individual people. Instead of aiming to produce *as much* happiness as possible, we should aim to make *individual people* as happy as possible.

But that leaves this question open: In making individual people as happy as possible, are we to maximize their average happiness, or their total happiness? Are we to make individual people as happy as possible per capita, or as happy as possible in the aggregate?

Jan Narveson and Rolf Sartorius interpret utilitarianism as the injunction to maximize the average happiness of the *actual population* of the world, past, present and future: Even given the ability, we have no obligation to produce more and more happy people, or to bring into existence a race capable of continuous ecstasy. Our obligation is just to make whatever population does or will *in fact exist* as happy per capita as *that* population can be.[2]

But this allows us to adopt the laissez-faire policy. For suppose that policy is adopted. Then the actual population of the world is as happy as *that population* can be (on the average *and* in the aggregate), because that population would not even have existed under the alternative policy, hence would not have been happier under the alternative policy.

Aggregate utilitarianism has the anomalous consequence that we should (not *may* but *should*) produce more and more happy people so long as that increases total happiness, even if average happiness plummets. Per-capita utilitarianism does not have this consequence. That is reason enough, one might contend, to prefer per-capita utilitarianism to the aggregate variety.

But per-capita utilitarianism has a fatal flaw. It conflicts with:

> PARETO OPTIMALITY   Suppose that A and B are among the feasible policy alternatives, that someone would be worse off if B were chosen than he would be if A were chosen, and that no one would be better off if B were chosen than he would be if A were chosen. Then B should not be chosen.

In fact, per-capita utilitarianism conflicts with an even weaker principle:

WEAK PARETO OPTIMALITY   Suppose that A and B are among the feasible policy alternatives, that everyone who would exist if A were chosen would enjoy some net benefit (some positive utility level) under A, that some people would exist under both alternatives and that *all* of them would be better off under A than under B. Then B should not be chosen.

To see how per-capita utilitarianism conflicts with Weak Pareto Optimality, consider this situation: There are just two feasible policy alternatives, P_1 and P_2. Those people who would exist under P_1 are the X-onians. They also would exist under P_2, as would the equally numerous Y-ites. The X-onians would be well off under P_1 and slightly better off under P_2. The Y-ites would obtain some net benefit under P_2, but not much, not enough to raise the P_2-average to the P_1-level; their lives would be worth living, but just barely.

According to Weak Pareto Optimality, since P_2 would make the X-onians better off and given the Y-ites some net benefit, P_1 should not be chosen, so P_2, the only other feasible alternative, *should* be chosen. But according to per-capita utilitarianism, since P_1 yields the greater average benefit, P_2 should *not* be chosen.

The conflict between per-capita utilitarianism and Weak Pareto Optimality is a symptom of a deeper difficulty with the whole utilitarian approach. Weak Pareto Optimality conflicts not only with per-capita utilitarianism but with:

MINIMUM BENEFICENCE   Suppose that A and B are among the feasible policy alternatives, that exactly the same people would exist under A as under B, and that A would yield a greater average utility, a greater total utility, a greater minimum share of utility and a more equal distribution of utility than B. Than B should not be chosen.

This captures the bare bones of the utilitarian intuition, broadly conceived. It does not require that all policy choices maximize social utility, in any sense of that phrase, although it does proscribe certain choices that could not be social-utility maxima by any criterion. Since it speaks only of policy alternatives that involve the same population, Minimum Beneficence is compatible with both per-capita and aggregate utilitarianism, and indeed with any principle for choosing among policy alternatives that involve different populations. Minimum Beneficence also is compatible with any modification of utilitarianism designed to give some weight to *distributive equality*. Although *distributive justice* is the same as distributive equality only in the unlikely case that people are equally deserving, Minimum Beneficence can be rendered compatible with any criterion of just desert by interpreting individual utility as *adjusted* utility – as 'raw' utility multiplied by a coefficient of desert.

Consider a new situation: There are three alternatives, P_1 and P_2 as before, and now P_3. Under P_3, as under P_2, the population would comprise the X-onians plus the Y-ites. The average benefit under P_3 would be slightly greater than under P_2. And the X-onians and Y-ites would be equally well off under P_3. The situation is depicted in the following bar graph:

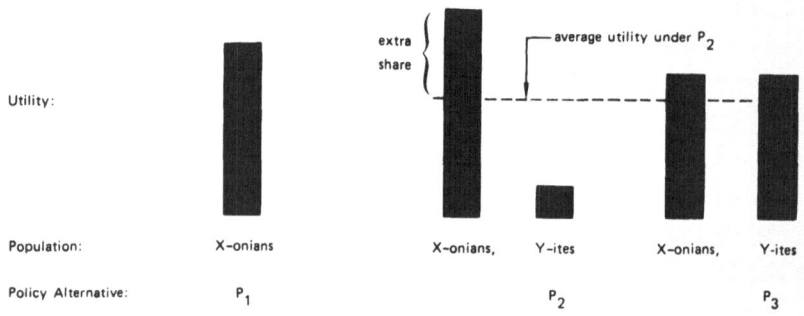

The population would be the same under P_3 as under P_2, and P_3 would yield a greater average utility, a greater total utility, a greater minimum share of utility and a more equal utility-distribution than P_2. According to Minimum Beneficence, then, P_2 should be rejected. But as before, Weak Pareto Optimality implies that P_1 should be rejected. And since everyone who would exist under P_1 would be worse off under P_3, Weak Pareto Optimality also implies that P_3 should be rejected. Taken together, then, Minimum Beneficence and Weak Pareto Optimality have the consequence that every feasible alternative should be rejected, which is impossible.

I conclude that we must reject Minimum Beneficence and therewith all versions of utilitarianism, broadly conceived, including all admixtures of utilitarianism with criteria of distributive justice.

The incompatibility of Minimum Beneficence and Weak Pareto Optimality can be stated formally and called an impossibility theorem. But nothing would be gained thereby.

For those utilitarians and fellow travelers who find Minimum Beneficence little more than a platitude, let me try both to dispel the air of paradox and to explain why it is that Minimum Beneficence (and utilitarianism in all its forms) came to grief. The X-onians either have a right to their extra share of utility under P_2 – their surplus over the average – or they do not. I think they do. But have it either way: If they do *not* have a right to the extra share, presumably the Y-ites do. But then, the availability of P_2 constitutes no reason to reject P_1 (as Pareto Optimality requires), since to adopt P_2 would be to bring about an unjust situation – one in which the Y-ites are deprived · of something to which they have a right. On the other hand, if (as I think) the X-onians do have a right to the extra share, then the availability of P_3 constitutes no reason to reject P_2 (Minimum Beneficence to the contrary not withstanding), since to reject P_2 in favor of P_3 would again be to bring about an unjust situation – one in which the X-onians are deprived of something to which they have a right. All forms of utilitarianism go wrong because they ignore established rights, or just claims, to specific benefits.

4. MEANWHILE, BACK AT THE PRESENT

Because P would impose costs on us and perhaps on some of our descendants, I don't see how P can possibly be justified unless not adopting P would make *someone* significantly worse off than he would be under P. But not adopting P would not make any of our *distant descendants* worse off than he would be under P. What of ourselves and our not-so-distant descendants?

If the benefits we and our near posterity would get from P are similar to those our distant descendants would get from P, then it is likely that some *less costly* policy would provide just us and our near posterity with the same benefits. And so long as such an alternative policy is available, we and our near posterity would be better off under it than under the more costly P.

Perhaps some of us would obtain great satisfaction from helping distant future generations to prosper. Surely, though, such selfless satisfaction is no part of one's own *welfare*.

But while present preferences for a flourishing future provide no basis for the contention that P would promote the welfare of the present generation, such preferences just might support a very different justification of P – one based on *fairness* rather than *welfare*.

Many of us want mankind to prosper and to continue to prosper on into the distant future. Among those who feel this way, each had rather that everyone contribute to this goal than that no one contribute. But apart from moral considerations, each had rather that everyone *else* contribute and *he not* contribute than that *everyone* contribute. It's like taxes: Each of us prefers everyone paying to no one paying. But apart from moral considerations and the fear of punishment, each prefers everyone but himself paying to everyone paying. In the case at hand: Many of us want our remote posterity to live in a clean, commodious, well-stocked world. Of those who feel this way, each had rather that everyone (himself included) bear a reasonable share of the cost of achieving these shared goals than that the goals not be achieved. But apart from moral considerations, each *most* prefers that everyone *else* bear the cost and he bear none of it.

Other things equal, however, it would be unfair for one who shared these goals not to bear a fair share of the cost of achieving them — just as it would be unfair for a citizen not to pay his taxes, for someone who liked clean public roads to throw trash out his car window, or for a member of a tug-o'-war team to feign pulling. In general:

> Other things equal, if every member of some group had rather that all members help achieve a certain shared objective than that none (or almost none) help, then it would be unfair to the others for one member not to help while the others (most of them, anyway) helped.

There are exceptions to this principle: The objective in question might be intrinsically illicit (lynching a Negro); its achievement might impose significant, undeserved external costs on someone (consumers, in the case of monopoly pricing); or a member of the group in question might merit an exemption (an elderly, infirm passenger in a stalled car that is being pushed by the other, more robust passengers). Still, there often is an obligation to observe the principle — to pull one's weight, to cooperate, not to be a free rider, not to take unfair advantage of one's fellows. One owes this obligation, when it exists, to those others who are helping to achieve the objective at issue.

So although *P* cannot be justified as promoting anyone's welfare, it is just conceivable that *P* can be justified as insuring fairness (enforcing cooperation, preventing free riders) among those who share the goal of a clean, commodious, well-stocked distant future.

University of Texas at Austin

APPENDIX

Brian Barry, John Harsanyi, Howard Margolis, William Galston and Gregory Kavka have raised interesting objections to my argument. The last three sent me written criticisms, each lucid, clever, insightful and deserving of publication – as well as wrong. Running through all these critical remarks were two main objections:

OBJECTION 1   Say we must give a windfall benefit to Mr X or Mr Y. We cannot give it to both. Neither has any special claim to it. No one but Messrs X and Y would be affected by either alternative. Mr X would benefit more if he got the benefit than Mr Y would if he got it. Surely we ought to give it to Mr X. The point is quite general, and it is applicable to the choice of long-range welfare policies: In deciding which of two potential beneficiaries to benefit, other things being equal, one should benefit him who would benefit more. So in the choice between adopting and not adopting P, we should adopt P (other things being equal), because that would benefit the $+ P$ population more than not adopting P would benefit the $- P$ population.

Reply   If we give the windfall benefit to Mr Y, then Mr X is worse off than he would otherwise have been, and he has grounds for the complaint that *he* ought to have gotten the benefit. But if we fail to adopt P, *no one* is worse off than he would otherwise have been, and *no one* has any grounds for complaint. Messrs X and Y have conflicting interests: We can benefit either one of them but not both. But there does not now exist and never will exist *two* distant future persons or groups with conflicting interests in the adoption vs non-adoption of P – *two* distant future persons or groups of whom we can benefit either one but not both.

OBJECTION 2   Before copulation, Bertha takes a drug for pleasurable effect, realizing that this will affect embryo-development, causing her to have a defective baby. It also affects meiosis and spermatozoon-selection, insuring that her baby is *different* from the one she'd have had if she had not taken the drug. By taking the drug, Bertha did not make her baby worse off than he'd otherwise have been, since otherwise he'd not have been. Surely, though, Bertha's act was wrong, and it was wrong because it harmed the baby.

Reply   Bertha's act was incredibly shortsighted, given her own interests. And it undoubtedly imposed significant costs on many other people. That adequately accounts for our feeling that Bertha did wrong and that the existence of the defective baby was a bad but avoidable consequence of her act. We need not suppose that she harmed or wronged *her baby* in any way — a paradoxical supposition, inasmuch as her baby would not otherwise have existed. A real counter-example to my implicit premise that ethically objectionable acts or policies must have victims would have to make it *clear and plausible* that the existence of the defective baby imposed no special costs on anyone. I challenge my critics to come up with such an example. I might add that I meant my argument to apply to individual family-planning decisions as well as public-policy decisions.

NOTES

[1] This paper partly overlaps my 'Obligations to posterity', in Brian Barry and Dick Sikora (eds.), Obligations to Future Generations (Temple University Press, Philadelphia, 1978).

[2] See Jan Narveson, 'Moral problems of population', The Monist 57 (1973), and Rolf Sartorius, Individual Conduct and Social Norms (Encino Dickenson, California, 1975), pp. 21–24.

STEVEN STRASNICK

MORAL STRUCTURES AND AXIOMATIC THEORY

ABSTRACT. Axiomatic decision theory has proven to be a valuable analytical tool in many disciplines, and in this paper I discuss its application to moral theory. The first part of the paper discusses the general structure of moral theory, and it argues that morality need not be identified with a particular moral principle. The concept of a moral framework is introduced, and a framework for use in analyzing issues of distributive justice is presented in the second section. The application of this framework is discussed in the paper's final section, and two different moral situations are analyzed. The utilitarian principle is argued to be appropriate for the first situation in which a scarce good is to be efficiently distributed, while Rawls' difference principle is claimed to be the correct one for the more abstract issue of basic institutional justice.

I

Axiomatic theory is a valuable analytical tool for investigating the validity of proposed decision criteria. Before it became widely adopted, decision criteria were often investigated in a piecemeal fashion. Proposed criteria would be tested against intuitive test cases and would either stand or fall on that basis. The axiomatic method replaced this procedure with a much more coherent approach, for now these intuitions could be systematically represented and their compatibility and logical interrelations examined. This enables us to achieve not only a better understanding of the fundamental differences between competing theories, but also a deeper insight into the relative validity of the intuitions themselves, as seemingly inocuous intuitions could be exposed as having unsavory consequences.

As it has been conducted in the last century, moral theory is a prime candidate for axiomatic analysis. More than any other decision-related theory perhaps, moral methodology depends on the investigation and propagation of counterexamples and intuitions. The literature is filled with the language of counterexample and counter-counterexample, with the result that the reader's head is often left spinning by the quick succession of intuition piled on intuition. The proponent of one ethical principle criticizes another by carefully constructing a hypothetical situation which can not be dealt with in an intuitively satisfying manner by the opposing principle, and the striken theorist replies in kind. And no theory, it seems, can escape the intuitive carnage.

Confronted with this literature, the skeptical reader may be tempted to conclude that all ethical principles are suspect, since all seem susceptible to the counterexamples of some imaginative critic or other. But such suspicions may be premature. It might indeed be the case that no ethical principle exists which can simultaneously satisfy the complete set of intuitions one has about morality. This, for example, was what Kenneth Arrow argued about social decision-making in his famous impossibility result.[1] But there is another explanation that may be offered for the seeming inability of any ethical principle to satisfactorily confront all proposed counterexamples. And that is that no single ethical principle should be able to deal with the complete range of ethical issues. Rather than seeing the existence of counterexamples as indicative of inherent contradictions in the nature of morality, we might view these counterexamples as fenceposts circumscribing the legitimate domains of different but not necessarily competing principles of ethics. Viewed in this manner, counterexamples may reveal important clues about the structure of moral theory – and not about its impossibility.

To understand the nature of this possibility, consider the case of the individual decision-maker. In his dealings with the world, the individual is confronted with a staggering array of decision-problems. Some decisions have to be made in the face of uncertainty or risk, others on the basis of incomplete information or ignorance, and still others against a reflective adversary. Obviously the individual wants to maximize his utility, but he needs a decision criterion that will tell him how to do this in each case. If the individual required a single criterion that he could employ in each situation, he would quickly be frustrated. And perhaps he would conclude that there was no rational theory of decision-making. On the other hand, if he listed the properties that rational decision-procedures should have and applied these to the particular circumstances of each situation, he would arrive at a set of different decision criteria for use in different situations.[2] Sometimes the expected utility criterion would be appropriate, other times perhaps a minimax criterion, and still others some version of the Arrow-Hurwicz criterion. The attempt to apply a critierion to the wrong type of situation would still result in counterexample. But this counterexample would not negate the rationality of the criterion as a whole. Instead, it would merely indicate the irrationality of a particular attempted application of an otherwise legitimate procedure.

Since we do not require the existence of a single decision-criterion as a pre-requisite for the rationality of individual decision-making, why do we view the nonexistence of a single all-inclusive ethical principle as indicative of the irrationality of morality? The answer, perhaps, lies in the history of moral thought. Traditionally, ethical theories seem to fall into two major camps: those arguing that there exist principles that are *constitutive* of the moral realm and those arguing that there do not.[3] Most classical theories fall into the former camp and devote their efforts to arguing that morality *just is* some principle (or perhaps combination of principles) that is to be applied in all cases. Well-known examples include the principle of greatest happiness or utilitarianism, Kant's categorical imperative, ethical egoism, the golden rule, etc.[4] Recent theories lean to the latter camp and argue that the ease with which all constitutive theories may be counterexampled is due to the fact that no principle is identical to the moral point of view. Rather, these theorists argue that ethics must be conducted on a case by case basis. Reasons in favor of certain courses of action must be weighed against opposing considerations, and the right action is that which has the greatest balance of reasons in its favor. No principles are assumed to exist which can perform this balancing. Rather, the right action is generated by an act of intuition.[5]

Faced with the above alternatives, the ethical theorist must either search for the mythical principle that can handle all possible counterexamples or resign himself to the inherent subjectivity of all ethical judgements. I suspect, however, that this choice between the constitutivist or the intuitionist is an arbitrary one, for there seems to be a way of reconciling these extremes if we regard morality as having a certain two-tiered structure to it. To be successful, this reconciliation must not only be capable of showing how the above alternatives would be generated, but must also demonstrate the possibility of a middle ground between these two extremes. It is this middle ground which we will hope to model with the tools of axiomatic decision-theory.

Let us consider, then, the following account of the structure of moral theory. Suppose, first of all, that actions and policies are morally right on account of their conformity with what we will call the *standards of moral rightness*. A standard identifies some property p such that actions having that property are right. For example, a utilitarian might hold that p is the property

of 'promoting the greatest happiness'. These standards are themselves determined by the *norms of moral relevancy*, which identify the types of factors
that count for and against the morality of a given action. An example of a
factor favoring the morality of an action is the fact that the action is impartial
and does not arbitrarily discriminate among individuals. If it were the case
that these norms always converged through some process of logical interaction
into a single standard of rightness, then some principle would be constitutive
of the nature of morality. This is, of course, the first alternative considered
above. On the other hand, if these norms never converged but instead
remained logically independent, we would be confronted with the second
alternative. The standard of rightness would simply be our intuition, and the
property p would be the property of 'having been picked by the intuition'.
But what of the third possibility? What if these norms sometimes converged
in certain situations and produced constitutive standards of morality appropriate for those situations only and not necessarily for any others? This
would be the middle ground to which we referred above – a ground I will
argue morality does in fact at times inhabit.

According to the conception of morality I wish to promote, morality is a
contextually sensitive discipline. Just as individual decision-theory is sensitive
to the context in which it is to be applied, with the consequence that different decision-criteria are generated in different situations, different constitutive standards of rightness will emerge in different moral situations as well.
The assumption is that there exists sets of relevancy norms that are indigen-
ous to specific areas of morality. Depending upon certain key situational
features of these areas, different standards of rightness will be generated by
these norms. This is not to claim that such sets of norms exist for all possible
moral situations, so that moral theory is complete. There may indeed exist
areas of morality that are not covered by these norms, areas for which moral
theory may be indeterminate. Rather, the claim is that there do exist certain
situations to which these special classes or norms, or *frameworks* as we will
call them, apply. In the second part of this paper, I will use the axiomatic
method to present a model of a type of ethical framework. This framework
will apply to a certain problem in moral theory, that of distributive justice,
and it will employ standard axioms of social choice theory to represent the
defining relevancy norms. In the third part of the paper, we shall see how
this framework may be applied to particular problems of distributive justice,

and we will consider the manner in which key features of different moral situations will generate different principles of justice.

II

Distributive justice is the branch of moral theory concerned with the evaluation of the moral status of the distribution of benefits and burdens, rights and duties, etc., produced by a society among its members. Like any branch of moral theory, distributive justice is a complex subject, and analyses that will work for one area of justice will not work for another. In the language of the first part of the paper, we might say that not only do we have to apply many different standards of rightness to the analysis of distributive questions, but many different frameworks as well. Because this is the case, areas for analysis have to be carefully circumscribed and simplifying assumptions made before we can wind our way down the web of morality to the point where frameworks become legitimate things for discussion. Ideally, frameworks will be specifiable in ways both sufficiently general to see how they may be applied in different areas within their domain of legitimacy and sufficiently specific to show how their analysis will lead to the determination of different standards of rightness.

We shall now develop a framework for analysis which will be called the *social welfare framework*.[6] This framework will make the assumption that the information relevant to the justice of a given distribution is specifiable in terms of the amount of some welfare-related value that each individual receives in this distribution. If x_i is the real-valued payoff function representing the amount of some welfare-related good that individual i receives in state x, each social state will be defined in terms of an ordered set of payoffs to individuals, with $x = (x_1, x_2, x_3, x_n)$. The preferences of individuals among these states will be determined by their payoffs, with individual i preferring x to y just in case x_i is greater than y_i. The problem confronting the social welfare framework will be to determine a standard of justice which will produce a social ordering R of the states S on the basis of some feature of the information about S that is identified as morally relevant by this standard. This account of the problem of distributive justice is, of course, purposely indeterminate, as it is intended to identify a certain general area within distributive justice for analysis. The area it picks out is just that set of problems within distributive justice that may be analyzed according to the above specifications without the loss of any morally relevant information.

Having specified the general features of the social welfare framework, we may next identify the axioms of social choice theory that we shall be employing to represent the relevancy norms of this framework. According to social choice theory, the problem of distributive justice is further refined, and it becomes one of determining how the competing preferences of individuals among the different social states are to be morally aggregated into a social preference. The axioms we assume will specify which features of these individual preferences will be morally relevant to this determination.

As originally formulated by Kenneth Arrow, social choice theory lead to paradox, as reasonable looking axioms modelling the moral point of view and rationality proved to be inconsistent. In previous work I have argued that the way to avoid Arrow's paradox is to require the social preference to be sensitive to information about the relative priority of individual preferences.[7] Individual i's preference for x over y has more priority that individual j's preference for y over x just in case state x would be socially preferred to y in the situation where these were the only two preferences. We shall be assuming that information about the priority of individual preferences is morally relevant to the social preference by imposing the following requirement, which we will call the *binariness axiom*: suppose we are considering two different situations in which society must choose between state x and y. If all individuals have the same preference in one situation as they do in the other and with the same degree of priority, then the same state must be socially preferred in each situation. As we shall see in the third part of this paper, it is this notion of preference priority that will be the contextually sensitive element of the social welfare framework. Depending upon what principles of preference priority are recognized as valid, different standards of justice will be generated by the framework. I will argue that certain features of distributive situations will point towards the moral relevancy of different principles of priority.

The second axiom of social choice we shall be assuming represents the impartiality aspect of the moral point of view. One of the requirements that is customarilly attributed to this point of view is the requirement that individuals who are identically situated with respect to relevant moral dimensions must have the same input into the social choice. This requirement will entail that any change in the names of individuals that leaves information about preference priority otherwise unchanged will leave the social preference unchanged as well. We shall also require that our standard of justice not discriminate between states on the basis of their names. Any change in the names of states over which individual preferences range will be reflected in the social preference. These two requirements define an anonymity and a neutrality condition and together establish the *impartiality axiom* of this framework.

The final axiom we shall impose on the social welfare framework captures and extends the benevolency aspect of the moral point of view. It has two components to it. The first component is known as the weak Pareto principle and requires that if all individuals are unanimous in their preferences regarding a state, society will prefer that state. The second component extends this unanimity requirement in the following way. Suppose that there is some way of dividing the society into a set of mutually exclusive and jointly exhaustive subsets of individuals with the property that the same state is socially preferred in each subset. Then this *extended unanimity* axiom will require that the social preference for the whole society be equivalent to the preference expressed in each of the separate subsets of individuals. Just as the weak Pareto principle requires that identical individual preferences be combinable in a consistent way into a social preference, this aspect of unanimity requires that identical social preferences be similarly combinable.

With the presentation of the three axioms of binariness, impartiality, and extended unanimity, we have completed our account of the relevancy norms of the social welfare framework. As we have described it, this framework represents a commitment to view complex moral phenomena as analyzable in terms of certain basic moral units. Clearly, not every problem in distributive justice will be amenable to analysis by this framework, and its range of application will have to be carefully limited by considerations of legitimacy, as we noted earlier. But there will be situations whose moral component can be effectively captured by the social welfare framework, and in the next section we shall consider two examples.

III

Because the social welfare framework has been defined in terms of certain indeterminate features, we must now consider the manner in which specification of these features will enable it to determine specific standards of justice. In particular, we have to identify the nature of the welfare-related value whose distribution is evaluated by the social welfare framework and determines the nature of the individual preferences among the different social states. For it is this dimension which has been left as the variable factor within this framework. Depending upon how this factor is specified in the application of the framework to particular problems of distributive justice, different principles of preference priority will be required, and these will in turn interact with the remaining norms to produce specific standards of distributive justice.

To see how this model of moral justification will function, we will now consider two different types of situations in which the social welfare framework may be applied. We shall argue that different principles of preference priority will be appropriate to each situation, with the result that different principles of justice will be entailed. The first case will be a situation in which scarce medical resources have to be distributed, and we shall see how features of this situation lead to the identification of a principle of priority which in turn entails the utilitarian standard of rightness. The second situation will model in a simplified manner the abstract approach to the problem of justice developed by John Rawls in his book *A Theory of Justice*.[8] This model will produce its own principle of priority and the subsequent determination of Rawls' difference principle as the appropriate standard of evaluation.

Suppose, for our first case, that society is threatened by a potential outbreak of a deadly new strain of flu. Fearing a crippling epidemic, a quantity of rare vaccine is obtained and readied for mass distribution. Unfortunately, there is not enough of the vaccine available to safely immunize all of the population. The more vaccine a person receives up to a certain threshold, the greater will be the possibility of his immunity to the disease. Different individuals will have different degrees of sensitivity to the vaccine depending upon identifiable physical traits, so that the same dosage will produce different degress of immunity in different individuals. Once the individual has contracted the disease, however, the results will be the same, independently of the physical condition of the individual or whether he was immunized or not. The same medical treatment will be required for any recipient of the disease, and the amount of physical suffering will be the same. Since the demand for the vaccine outsrips its supply, a decision will have to be made concerning a policy of distribution.

From the standpoint of the social welfare framework, this situation will be representable in the following manner. Alternative social states will appear as different institutional policies of vaccine distribution. The individual's welfare-related payoff in each state will refer to the probability of immunity to the disease produced in him by the dosage he receives under the immunization program of that alternative. The problem confronting the framework is to define a standard which will rank order the different immunization programs on the basis of this immunity-related information. As we have developed the framework, this standard will be determined if a principle for weighing the priority of different individual preferences can be defined. Since individual preferences are determined by the degree of immunity the individual receives in each social state, the principle of priority must evaluate these preferences on the basis of this information.

What kind of principle of priority would be morally appropriate to this situation? The answer to this question, I would claim, depends on the nature of society's purpose in distributing the vaccine. Since the crucial consideration from the standpoint of society is just to determine which distribution of the vaccine will produce the most effective utilization of a scarce good, the nature of each individual's priority from a social standpoint should be a function of this factor. Thus, it seems reasonable to view the individual's degree of effectiveness in utilizing the vaccine as the socially relevant feature of his situation which determines his preference's relative priority to that of other competing preferences. Since the individual's degree of effectiveness in utilizing the vaccine in state x versus state y where $x_i > y_i$ is simply $x_i - y_i$, we may obtain the following general principle of priority for any x and y: for xP_iy and yP_jx, if $x_i - y_i = y_j - x_j$, then xP_iy equals yP_jx in priority. This priority rule, which we will call the *effectiveness* priority principle, can be shown to logically entail the utilitarian standard of rightness in conjunction with the axioms of the social welfare framework.[9] One state will be socially preferred to another just in case there is a greater degree of effective utilization in that state as measured by the sum of individual utilizations than in another.

The effectiveness priority principle that we have introduced seems to be a principle that should have wide application in many kinds of distributive situations in which the efficient distribution of a scarce good is at stake. And that in turn means that the utilitarian standard is the appropriate one to use in a wide range of cases. While that is probably the case, we should be careful about concluding from that possibility that the utilitarian standard is the only correct one for use in evaluating distributive questions of scarce goods. For there might be cases in which the individual's effectiveness in utilizing the social good is not an appropriate consideration. In the following discussion I will present an example of a situation in which this will be the case.

Accordingly, suppose that instead of being concerned with the distribution of scarce welfare-promoting goods, we are investigating the issue of the basic institutional design of society. Following Rawls' account, we are to determine the nature of the institutions through which society assigns systems of rights and liberties, opportunities and offices, wealth and power, etc., to its members. These goods will be of social origin and their contribution to individual welfare will be measured in terms of what Rawls calls an *index of primary social goods*. This index does not represent the individual's ability to make use of these goods in the conduct of his life, but is instead and objective indicator of the amount of these goods received by the individual within the institutional structure of society.[10] These goods are to be understood as general all-purpose means to the individual's pursuit of his ends, and his preferences among the different social structures will be defined in terms of the payoffs of primary goods received from these structures.

To investigate the nature of preference priority in this situation, let us begin with a simple two-person case. Suppose individuals 1 and 2 are each in a situation of equality with respect to the socially determined index of primary goods. If institutional structure x is realized, individual 1's index will be increased due to a more favorable allocation of social goods while the index of individual 2 would remain the same. If y is realized, the opposite will be the case. Now how do we evaluate the question of which state should be morally preferred? Suppose it were the case that individual 1 would receive a higher index of primary goods in x than individual 2 would receive in y. Would this difference in allocation be morally relevant to the determination of their respective priorities of preference? If it were the case that this higher index of primary goods was due to the fact that individual 1 could more effectively utilize an allocation of these goods than individual 2, then perhaps we would be justified in concluding that individual 1's preference had a higher priority. Our reason for doing so would stem from the same kind of efficiency considerations that motivated our evaluation of the vaccine case that we discussed above. However, in the present case, considerations of efficiency are not involved. This is because the index of primary goods is not a measure of the individual's subjective capacity of utilization, but is instead a simple objective measure of the amount of these goods that he receives from society. Since the index of primary goods does not furnish us with any moral basis for differentiating between the claims of individuals 1 and 2 for primary goods in this situation, we would have no basis for socially preferring state x over state y.

If this argument is right, then individual 1's preference must be regarded as having the same priority as that of individual 2, even if it were the case that individual 1 would receive a higher allocation of primary goods in his preferred state than would 2. In such a situation of initial equality, each individual must have an equal claim to further allocations of primary goods. For this two-person case, therefore, we have obtained the following principle of preference priority, which will be called the *equity* principle: for xP_1y and yP_2x, if $y_1 = x_2$, then xP_1y equals yP_2x in priority. And it can be shown that this equity principle, in conjunction with the axioms of the social welfare framework, will logically entail Rawls' difference principle, so that the social preference must always be equivalent to the preference of the individual who would be left worst-off if his preference were not satisfied.[11] This principle is strongly egalitarian, for it sanctions an individual's being left worst-off within a given social arrangement *only* when any other institutional arrangement would leave someone even more worse-off.

With this conclusion, we have completed our examination of the implications of the social welfare framework for certain problems in distributive justice. In situations where society is legitimately concerned with the efficient distribution of scarce goods, we have argued that the ability of the individual to effectively utilize these goods is the morally relevant factor in the evaluation of the priority of his preference. This factor is not relevant, however, if we accept Rawls' account, in the more abstract situation of basic institutional justice, and considerations of equity become overriding in the evaluation of individual priority. Depending upon which principle of priority is appropriate, different standards of justice will emerge, so that we have seen an illustration of the sense in which a moral framework can account for the existence of different moral principles within the same general moral realm. Further analysis will have to determine if the concept of a framework is a useful analytical tool in areas of morality other than distributive justice. It should be clear, however, that there does exist a viable alternative to the view that the point of view of morality can only be represented by one all inclusive moral principle. That one principle should lord over all the realm of morality strikes me as absurd as the belief that one individual should always determine the social preference independently of the wishes of everyone else. Conceptual dictators are no different in kind than their human counterparts.

Stanford University

NOTES

[1] See Arrow's: Social Choice and Individual Values 2nd ed. (New York, 1963).

[2] A good example of this kind of treatment is found in R. D. Luce and H. Raiffa, Games and Decisions (New York, 1957), ch. 13.

[3] This distinction is discussed in B. Rosen, Strategies of Ethics (Boston, 1978), ch. 5.

[4] For a classic discussion of these ethical theories, see H. Sidgwick, Methods of Ethics (New York, 1966).

[5] Examples of theorists holding this view are E. F. Carritt, Thoery of Morals (Oxford, 1928) and H. A. Prichard, Moral Obligations (Oxford, 1949).

[6] This framework and its axioms are discussed in more detail in my 'Preference priority and the maximization of social welfare', (Ph.D. thesis, Harvard University, 1975).

[7] See my 'The problem of social choice: Arrow to Rawls', Philosophy and Public Affairs 5 (1976), pp. 241–273, and 'Ordinality and the spirit of the justified dictator', Social Research 44 (1977), pp. 668–690.

[8] See especially Rawls, A Theory of Justice (Cambridge, 1971), part I.

[9] This proof is a simple extension of that found in my 'Preference priority and the maximization of social welfare', ch. 5.

[10] Rawls discusses the notion of primary goods in: A Theory of Justice, section 15.

[11] See my 'Social choice and the derivation of Rawls' difference principle', Journal of Philosophy 73 (1976), pp. 85–94. The argument presented in this paper is somewhat similar to that presented above, though it is stronger. For I argue that the nature of Rawls' argument precludes giving individual 1's preference a higher priority *even if* this index of primary goods measures the individual's capacity to utilize these goods.

DONALD WITTMAN

A DIAGRAMMATIC EXPOSITION OF JUSTICE*

ABSTRACT. Several different analytical theories of justice are analyzed in the context of a uniform diagrammatic exposition. The diagrams make use of ordinary economic tools of analysis – production possibility frontiers and indifference curves. Additionally, numerous propositions are set forth which establish either the critical differences between alternative theories, or the conditions under which different theories yield identical results.

In this essay, several ostensibly unrelated analytical theories of justice are analyzed within the context of a concise and uniform diagrammatic exposition. The diagrams make use of ordinary economic tools of analysis – production possibility frontiers and indifference curves. The geometry involved in these diagrams has two advantages over mathematical logic (the typical mathematical mode of discourse on justice). The concepts are more accessible to those who are not skilled in symbolic logic. Furthermore, many new insights can be gained and interrelationships discovered by seeing the same problem formulated from a different mathematical perspective (in this case geometry). Thus the paper satisfies two objectives: to provide new research results for those active in the field, and to serve as a pedagogical device for those who are not. Throughout the paper numerous propositions are presented which show either the critical differences between alternative theories of justice, or the conditions under which different theories yield identical results.

The analysis will focus on the following dimensions of justice:

(1) Choice of origin for utility measurement;

(2) Choice of justice indifference curve;

(3) Method of measurement of utility tradeoff frontier;

(4) Interpersonal comparisons (or no interpersonal comparisons) of utility;

(5) Cardinal, ordinal or semi-cardinal utility functions; and

(6) Whether a social contract is made with full information regarding objective position and subjective preferences, with partial information, or with complete ignorance.

Theory and Decision 11 (1979) 207–237.   0040–5833/79/0112–0207 $02.10

In Section I, the baseline measure is zero utility. Most of the justice criteria are introduced in this section and a number of relationships are derived. In Section II, a different baseline is used, that of the status quo. Social contract theories make use of the status quo in that a minimum requirement for a social contract is that all the participants are at least as well off after the contract as before. In Section III, the social contract is make in a veil of ignorance. No one knows who he will be after the contract is completed. That is, Person A may end up being Person B by having B's objective circumstances as well as B's subjective preferences (utility function). In Section IV, partial empathy is considered. In this case Person A can put himself in Person B's objective circumstances but Person A maintains his own utility function. Social contracts are analyzed under this partial veil of ignorance (partial because A knows whose utility function he will have after the contract is completed — his own, but not whose objective circumstances he will acquire). The concept of minimal envy, which also makes use of partial empathy, is developed in this section. In Section V, equality of income is analyzed.

I. INTERPERSONAL COMPARISONS OF UTILITY – FROM ZERO UTILITY

A. *Utility Tradeoff Frontier and Pareto Optimality*

It is useful to start off with a utility (happiness) tradeoff frontier (which is analogous to the production possibility curve in economics).[1] In Figure 1, any points within or on the curve are feasible. The shape of the utility frontier implies that once on the frontier, not only does an increase in A's utility result in a decrease in B's utility, but also that additional units of A's utility

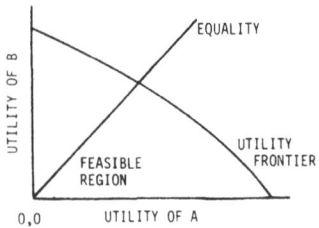

Fig. 1. Utility tradeoff frontier. Note that the curve need not stop at the axes.

can only be achieved at an increasing loss of B's utility (other configurations of the utility frontier will be considered in later figures).

A Pareto optimal point is one at which one person's utility cannot be increased without decreasing another person's utility. Pareto optimality is equivalent to economic efficiency and in this case is equivalent to the utility tradeoff frontier line. Pareto optimality is an ordinal concept (i.e., it measures only greater, lesser or equal) and does not involve interpersonal comparisons of utility (i.e., one person cannot be said to be happier than another person).[2] Thus, all the points on any utility tradeoff frontier line which slopes downwards from left to right are pareto optimal.

The 45 degree line is a line of equal utility for A and B. It should be noted that this need not imply equal money income nor need equal money income imply equal utility. It is also important to note that equality and efficiency are not mutually exclusive for perfect equality is one point on the efficiency frontier and efficiency is one point on the equality frontier (the 45° line).

B. Maximin Criterion

Justice measures, which involve weighting of the utility functions, will be considered first. Later these will be shown to be related to notions of justice which incorporate the concept of a veil of ignorance.

The maximin criterion of justice weights only the utility of the worst off member of society (measurement being made from the zero utility baseline).[3] It thus makes ordinal interpersonal comparisons of utility (A has more or less utility than B). If B's utility is less than A's, then an increase in A's utility is

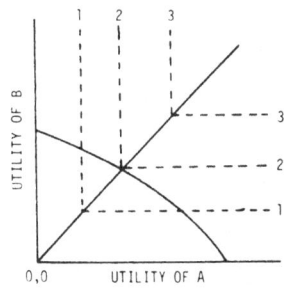

Fig. 2. Maximin criterion. This criterion has ordinal interpersonal comparison of utility. – – – Maximin indifference curves.

not more desirable according to the maximin criterion. This can be seen in Figure 2 where a series of justice indifference curves based on the maximin criterion are drawn. 2, 2 is preferred to 1, 1 because the minimum is higher. Points above 1, 1 are preferred to points on 1, 1 which are preferred to points below 1, 1. All the points on 1, 1 are indifferent to each other according to the maximin criterion as the minimum utility is the same. In mathematical terms J is a monotonically increasing function of the minimum $U_i(\cdot)$ $i = 1$, 2, ... N where J is justice and U_i is the utility of the ith person (for mathematical simplicity the mathematical notation will refer to person 1, 2, etc. and not person A, B, etc.).

C. Sum of Utilities or Utilitarian Approach

The sum of utilities criterion (often referred to as the utilitarian approach — a somewhat misleading phrase) weights the utility of each person equally (like all justice measures in this section the baseline is zero utility).[4] It thus involves cardinal interpersonal comparisons of utility.[5] If A has 10 units of utility and B has 0 units of utility, this is equivalent to 5 units of utility for A and 5 units of utility for B which is equivalent to 10 units of utility for B and 0 units of utility for A. For the sum of utilities criterion the justice indifference curves are straight lines forming a 45° angle with the horizontal and vertical axes (see Figure 3). Thus the dispute between maximin and utilitarian approaches to justice centers on the shape of the justice indifference curves. In mathematical symbols J is a monotonically increasing function of $\Sigma_{i=1}^{N} U_i(\cdot)$.

The problems with both the maximin criterion and the sum of utilities

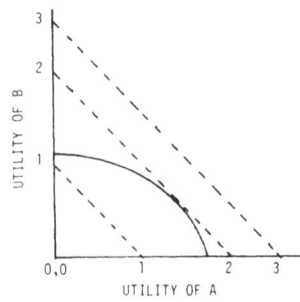

Fig. 3. 'Utilitarian' or sum of utilities criterion. This criterion has cardinal interpersonal comparison of utility. – – – Utilitarian indifference curves.

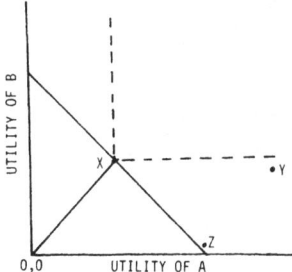

Fig. 4. Problems with the two criteria. X is at the intersection; Z is preferred to X by the sum of utilities criterion (——); X is preferred to Y by the maximin criterion (– – –).

criterion are very apparent. The maximin criterion does not weigh improvements of those who are not least well off. Thus by the maximin criterion, a situation where A is very miserable and B is only mildly so is preferred to a situation where A is slightly more miserable (he has one less handkerchief to cry on) and B is in Nirvana (i.e., X is preferred to Y in Figure 4). On the other hand, the sum of utilities criterion would prefer A to be very happy and B to be miserable, rather than both to be just a little less than half way between very happy and miserable (i.e., Z is preferred to X in Figure 4).[6]

D. Compromise Criteria

In many ways these two criteria, maximin and sum of utilities, can be seen as the limiting cases.[7] As suggested by the above discussion most people would have justice indifference curves somewhere in between (see Figure 5). There are a variety of compromise justice functions. Examples include: the

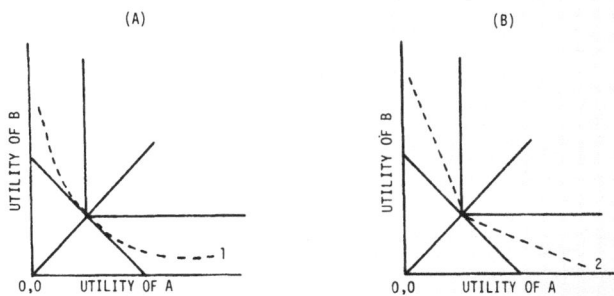

Fig. 5. Compromise criteria. – – – compromise criterion indifference curve.

summation of the square roots of the utility functions $(\Sigma_{i=1}^{1}[U_i(\cdot)]^{1/2})$, and the product of the utilities or equivalently the summation of the log of the utilities $(\Pi_{i=1}^{N} U_i(\cdot)$ or $\Sigma_{i=1}^{N}$ Log $U_i(\cdot)$. This approach will be developed further in a later section.) These justice functions would produce justice indifference curves somewhat similar in shape to the compromise criterion indifference curve in Figure 5a. Another approach is to make a new justice indifference curve from the weighted function of the maximin and sum of utilities justice function. In Figure 5b, compromise criterion 2 is the following weighted sum

$$\frac{2 \text{ Maximin} + 1 \text{ sum of Utilities}}{3}.$$

(For $U_A > U_B$, the function is as follows:

$$\frac{2U_B + 1[U_A + U_B]}{3} = \frac{U_A + 3U_B}{3}.$$

This justice indifference curve is identical to a welfare function derived from the Gini coefficient.[8] For $U_A > U_B$ the Gini coefficient of inequality is

$$1 - \frac{1}{2} + \frac{2}{4\bar{\bar{U}}}[U_A + 2U_B]$$

where

$$\bar{U} = \frac{U_A + U_B}{2}.$$

A homothetic social welfare function $W(g)$ is created by multiplying average utility times the [Gini coefficient minus one].

$$W(g) = -\frac{\bar{U}}{2} + \frac{2\bar{U}}{4\bar{\bar{U}}}[U_A + 2U_B],$$

$$= -\frac{U_A + U_B}{4} + \frac{2U_A + 4U_B}{4},$$

$$= \frac{U_A + 3U_B}{4}.$$

This creates the same set of indifference curves as

$$\frac{U_A + 3U_B}{3}.$$

E. Egalitarian Approaches

There are a variety of egalitarian measures. They are extreme measures, valuing only equality and not utility of the participants. For all egalitarian criteria the highest indifference curve is the 45° line from the origin. Egalitarian criteria differ among themselves when the indifference curves are not at the 45° line. We will explicitly consider two criteria: the absolute egalitarian criteria and the Gini coefficient. The former will be shown to be diametrically opposed to the sum of utilities criterion while the latter will be shown to be diametrically opposed to the product of the utilities criterion. Two criteria are said to be diametrically opposed if for every point U_A, U_B, the slope of the justice indifference curve of the first criterion is *minus* the slope of the indifference curve of the second justice criterion.

We consider the absolute egalitarian measure first. If absolute differences are the measure of inequality then there are pairs of indifference curves parallel to the 45° line. The closer to the 45° line, the better according to this criterion (see Figure 6).[9] As can be seen, this criterion results in indifference curves perpendicular to the sum of utilities approach. It is thus diametrically opposed to the sum of utilities criterion. Again this criterion involves ordinal interpersonal comparisons of utility. In mathematical symbols it is a monotonically decreasing function of

$$\sum_{i=1}^{N} |U_i(\cdot) - \bar{U}(\cdot)|$$

where | | stands for absolute value and $\bar{U}$ stands for average utility.

We next consider the Gini coefficient of equality (which differs from the

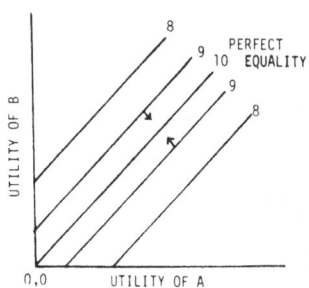

Fig. 6. Egalitarian criterion. The arrows point towards the preferred direction.

welfare function based on the Gini coefficient as shown earlier). The Gini coefficient indifference curves are diametrically opposed to the product of the utilities criterion.

Proof. Along an indifference curve the measure is constant. Therefore if we take a total differential and set it equal to zero we can derive the slope of the indifference curve at any point.

For $U_A > U_B$

$$G = 1 - \frac{1}{2} + \frac{[U_A + 2U_B]}{U_A + U_B},$$

$$dG = \left[\frac{1}{U_A + U_B} - \frac{U_A + 2U_B}{[U_A + U_B]^2}\right] dU_A$$

$$+ \left[\frac{2}{U_A + U_B} - \frac{U_A + 2U_B}{[U_A + U_B]^2}\right] dU_B = 0.$$

Combining terms

$$= \frac{U_A + U_B - U_A - 2U_B}{[U_A + U_B]^2} dU_A$$

$$+ \frac{2U_A + 2U_B - U_A - 2U_B}{[U_A + U_B]^2} dU_B = 0,$$

$$= \frac{-U_B}{[U_A + U_B]^2} dU_A + \frac{U_A}{[U_A - U_B]^2} dU_B = 0$$

$$\text{or} \quad \frac{dU_A}{dU_B} = \frac{U_A}{U_B}.$$

For the multiplication of utilities criterion denoted by N (for Nash) $N = U_A U_B$.

To find the slope of the indifference curve we take the total differential and set equal to zero.

$$dN = U_A dU_B + U_B dU_A = 0 \quad \text{or} \quad \frac{dU_A}{dU_B} = \frac{-U_A}{U_B}.$$

Thus the slope of the Gini coefficient indifference curve is minus the slope of the multiplication of utilities (or Nash) criterion.

F. Identical Results and Similar Properties

The nature of individual preferences and production of outcomes need not result in a smooth and connected utility feasibility set.[10] In Figure 7 the feasible utility possibilities are shaded and the various criteria applied. As can be seen the criteria can greatly diverge.

I will now present sufficient conditions for all the criteria to yield identical results.

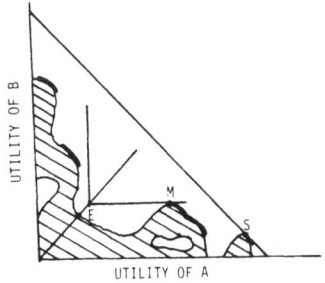

Fig. 7. A comparison of the justice criteria when the feasible set is irregular. Feasible set – Shaded in area. Maximin criterion – M. Sum of utilities criterion – S. Egalitarian criterion – E. Actually E is the whole line segment from 0 to E, but if we have a lexicographic ordering (Choose the highest equality indifference curve first; then choose the farthest feasible point away from the origin on this indifference curve), then we have the unique point E. Pareto optimal points – Heavy dark line.

PROPOSITION 1. If the set of pareto optimal points includes a point on the 45° equilibrium line, then the maximin criterion satisfies the egalitarian criterion.

Proof. Starting from the intersection of the 45° line with the pareto optimal line (point X) we cannot improve the utility of one person without decreasing the utility of the other. If we could, point X would not be a pareto optimal point.

PROPOSITION 2. If the set of pareto optimal points includes the 45° egalitarian line, is symmetric around the 45° line and is concave (i.e., bowed out from the origin), then the maximin criterion satisfies the sum of utilities criterion and the egalitarian criterion.[11]

Proof. The second part follows from the first proposition. Concavity and

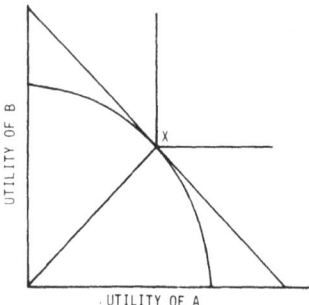

Fig. 8. Equivalent results. When the utility tradeoff frontier is symmetric and bowed out then $X = E = M = S$.

symmetry means that at the point of equality the sum of the utilities cannot be less than the sum of utilities at any other feasible point. By symmetry and concavity, the sum of utilities indifference curve is tangent above the pareto optimal curve at the point of equality (see Figure 8).

PROPOSITION 3. Let $M = $ maximin criterion point, $S = $ sum of utilities point and EE be the egalitarian line, then the distance from M to EE is not greater than the distance from S to EE.[12]

Proof. Assume the contrary that S is closer to EE. If the smaller utility is larger in S, this violates M which maximizes the smallest utility. If the smaller utility is not larger in S, then for S to be closer to EE the larger utility must be smaller in S, but this violates S which maximizes the sum of utilities.

The sum of utilities approach, as well as any other criterion with indifference curves that slope downwards from left to right and that are symmetrical (have mirror images) around the 45° line, have three desirable qualities: pareto optimality, anonymity and independence of irrelevant alternatives. The maximin approach has anonymity and independence of irrelevant alternatives. If we make the maximin criterion lexicographic by first choosing the indifference curve where the minimum utility is highest and *then* choosing a feasible point on this curve which maximizes the utility of the other person, maximin is also pareto optimal (this is known as leximin).

The sum of utilities is pareto optimal because a downward sloping indifference curve means that two positions are indifferent to each other only if one person's utility is higher and the other person's utility is lower. One

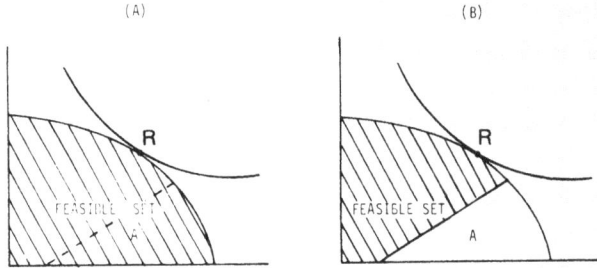

Fig. 9. Independence of irrelevant alternatives. The indifference curve is independent of the feasible set.

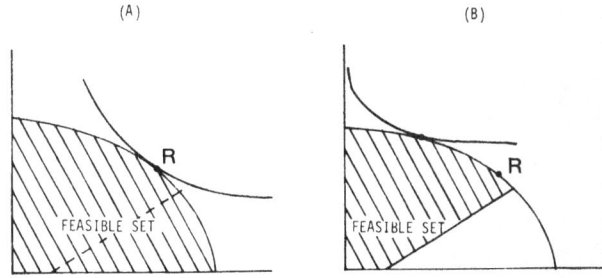

Fig. 10. Not independent of irrelevant alternatives. The indifference curve is dependent on the feasible set.

indifference curve is higher than another indifference curve only if it contains some position which has higher utility for both individuals than some position on the lower indifference curve.

Anonymity (sometimes called symmetry) means that if we relabel the axes we would make the same choice: that is, there is nothing special about being A (or B) so that if we relabel, A should now receive what B received before the relabeling and B should receive what A received before. This symmetry towards the participants is clearly seen in the symmetry of the indifference curves around the 45° line.

Independence from irrelevant alternatives means that if an allocation R is chosen when A is a feasible alternative, R will still be chosen when A is no longer an alternative. This quality can readily be seen in Figure 9 when the highest feasible indifference curve is attained at R. Eliminating some other feasible points does not alter the fact that the highest feasible indifference curve is reached at R. If the shape of the indifference curves depended on the feasible set as in Figure 10 they would no longer be independent of irrelevant alternatives.

II. THE STATUS QUO AS A BASELINE AND SOCIAL CONTRACT THEORIES

A. *The Status Quo as a Baseline*

In the theories considered thus far the baseline measurement is zero utility. Another approach is to treat the status quo (S_1, S_2) as the point of origin. The just outcome is then measured in *additional* utilities from the status quo. For example, the maximin criterion with the status quo as a baseline (M^*) now maximizes the minimum additional utility (see Figure 11).

An important class of theories which make use of the status quo are social contract theories.[13] Under a social contract the members of the collectively voluntarily enter into a contract. This unanimous agreement can only be achieved if everyone under the contract is at least as well off as under the status quo of no agreement. Thus in Figure 11 the original formulation of the maximin criterion (M) with 0, 0 as the origin cannot be arrived at via a unanimous agreement as A would be worse off than the status quo point S_1, S_2.

In some ways this might seem to be unjust for no extra consideration is given to B whose status quo point was low to begin with. I.e., there is no attempt to make up for B's initial weak condition. Thus, if the status quo point has B being very poor and unhappy and A being very rich and happy, B

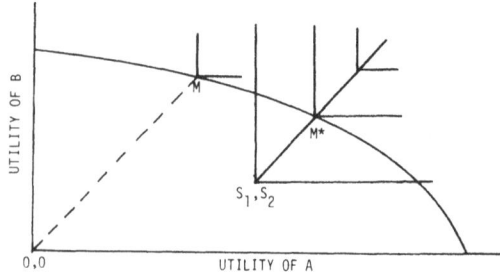

Fig. 11. The status quo. Point M maximizies minimum utility. Thus A would be worse off than at the status quo. In contrast, M^* maximizes minimum additional utility.

will not be compensated for being poor. In fact, when only income is allo-cated, if there is decreasing marginal utility of income, most of the extra income accruing from the social contract will go to the rich man as equal marginal increments of utility imply more additional income to A (the rich man who gains less utility per unit of income) than to B, the poor man.[14]

For this reason, most people who discuss social contract theories assume that the participants start from equal positions. The 'State of Nature' is typically uniform for all. Diagramatically, this means that the status quo is somewhere on the 45° ray from the origin (and consequently $M = M^*$).

B. Product of Utilities or Nash Criterion

There is, however, another way around the problem of an unequal status quo – we can turn the vice into a virtue. Using the status quo as a baseline ignores differences in initial utilities between the two participants so why not choose a method which makes no interpersonal comparisons of utility whatsoever. After all, it is rather difficult to say that A is happier than B; so maybe we should try to avoid interpersonal comparisons of utility altogether. (Another solution to the problem of interpersonal comparisons of utility is found in Part IV – partial empathy.) In order to understand how this might be done it is useful to know how utility indices are constructed in the first place.

We may observe that person A, when given a choice between a pound of steak and a pound of hamburger (at the same price), will choose steak; when given a choice between hamburger and eggs (again at the same price per pound) will choose hamburger; furthermore, A, when given a choice between the sure thing of hamburger and a lottery which yields steak P percent of the time and eggs $[1 - P]$ percent of the time, will always choose the lottery if P is greater than 0.8 and will always choose hamburger if P is less than 0.8. Person A is thus indifferent between the sure thing of hamburger and the lottery giving steak 80% of the time and eggs 20% of the time. This prefer-ence relationship can be represented by a utility function which is unique up to a linear transformation if we assume that A maximizes his expected utility. This is a von-Neumann-Morgenstern utility function. We can assign 80 units of utility for hamburger, 100 units of utility for steak and 0 units for egg as $100\% \cdot 80 = 80\% \cdot 100 + 20\% \cdot 0$. Any positive linear transformation will also satisfy the equation (e.g., $H = 8$, $S = 10$, $E = 0$; $H = 9$, $S = 11$, $E = 1$). It would be unfortunate if a justice rule depended on an arbitrary decision between representing the utility function as $H = 9, S = 11, E = 1$ or $H = 80$, $S = 100$, $E = 0$.[15] Looking at the issue from another perspective, it would be strange that our decision would change just because we change our units of measurement (after all, if we decide to buy two feet of cloth, we should also decide to buy 24 inches of cloth).

Thus, we might want a justice criterion to have the following quality: that if the justice criterion leads to A_1 units of utility for A and B_1 units of utility for B, a positive linear transformation of A's utility function will lead to a new A_1^* which is the same linear transformation of A_1 while B_1 will remain the same. In this way the choice of goods and services will remain the same. Since a linear transformation can make A's utility larger or smaller than B's or make differences between A's utilities larger or smaller than B's and yet not affect the just amount of B's utility, we do not have interpersonal comparison of utility.

This quality (outcome not affected by a linear transformation of utility scales) along with the three qualities previously mentioned, pareto optimality, anonymity and independence of irrelevant alternatives, is only satisfied by one justice criterion. This criterion is known as the Nash solution to the bargaining game and maximizes the multiplication of the increments to utility from the status quo (equivalently, the sum of the logarithms of the utilities are maximized).[16] The indifference curves are rectangular hyperbolas centered at the status quo (see Figure 12). Rectangular hyperbolas form rectangles of equal area. Thus, if we start with four 'reasonable' qualities that any justice criterion should have, we are led to one particular criterion. While multiplication of (additional) utilities may seem to be intuitively less satisfying than addition of (additional) utilities, it uniquely satisfies there four qualities.

One can attack the Nash solution on two grounds: (1) the desirability of these four qualities and (2) the measuring of the status quo. Typically, independence of irrelevant alternatives is attacked as some people feel that non-winning alternatives should still effect the optimal choice. The meaning of status quo is also often attacked. The criticisms apply to almost any social contract theory. It is very hard to determine what we mean by status quo. If we believe the contract is made in a state of nature, does it mean that the participants are in isolation, or that they are in contact but do not use threats or that they are in contact and use threats?[17] Is it really appropriate (especially in the twentieth century) to say the 'state of nature' is the status

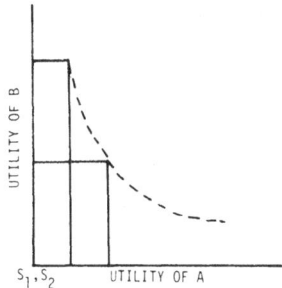

Fig. 12. Product of utilities indifference curve. The Nash approach has a rectangular hyperbola for an indifference curve. All rectangles are of equal size. There is no interpersonal comparison of utility.

quo? Perhaps the present state of affairs should be considered the status quo (perhaps via immigration the present state of affairs is the unanimous agreement; or perhaps there is no unanimous agreement as to the status quo). Furthermore, social contract therories put a heavy emphasis on the status quo and why should a just theory say the status quo is the appropriate or just baseline.

III. SOCIAL CONTRACTS IN A COMPLETE VEIL OF IGNORANCE

One approach towards justice is to ask what rules or outcomes (e.g., utility or income distribution) the participants would choose if the participants did not know which person they would be once the rules were instituted. In this situation, the person would 'selfishly' maximize his anticipated utility without knowing who he would be. The outcome is just because the choice is made ex ante without knowledge of who will occupy particular positions. If people voluntarily agree to these rules, we have a social contract made in ignorance, where all participants prefer being part of the contract rather than remaining at the status quo (Section II can be viewed as a social contract made with knowledge of who the individuals will be). An alternative to the status quo as the origin of measurement is an 'enforced social contract' (a somewhat contradictory concept) where the origin is zero utility.

In this section we assume that the social contract is made in a compete veil of ignorance. People make a contract not knowing which position, either in terms of objective circumstances or subjective preferences, they will be in after the contract is made. That is, *A* may end up being in person *B*'s objective circumstances with *B*'s utility function after the contract is made. We will show that under a complete veil of ignorance the difference between the maximin and sum of utilities (utilitarian) criteria is based on the difference between risk and uncertainty. With risk the probabilities of particular states of the world are known. The appropriate behavior under risk is to maximize expected utility. With uncertainty, probabilities are unknown as the outcome is determined in an *N*-person game. For zero-sum games, the appropriate strategy is a maximin strategy.

A. Sum of Utilities (Utilitarian) Criterion under Risk

If individuals in a veil of ignorance can assess the probability of being particular individuals after the veil has lifted and can assess the probability of various events occurring to these particular individuals, then there is decision making under risk. It is quite plausible that individuals can predict the probability of being particular individuals for if there are *N* people, the person has probability $1/N$ of being any particular person. However, it is less plausible that individuals can assess probabilities of all states of the world.

With risk if a person obeys the von-Neuman Morgenstern utility axioms, self interest in a veil of ignorance will result in expected utility maximization.[18] The choice will then be that point which maximizes the sum of utilities, for the person has probability $1/N$ of being any particular person. His expected utility (*EU*) is then the intersection of the 45° equality line and the maximal sum of utilities (see Figure 13). The choice will always be at *S* (sum of utilities) but since the person will be *A* half the time and *B* half the time, the person can expect *EU*. Thus, in the veil of ignorance, everyone has the same *ex ante* (before they know which position they will be in) utility but different *ex post* utilities when the veil is lifted. It is important to note that expected utility incorporates distaste for risk. For example, a person may be indifferent to a lottery which offers him a fifty percent chance of $10 and a fifty percent chance of $1000 to a sure thing of $100. This may be characterized by a utility function which places 0 'utilities' for $10, 5 'utilities' for $100, and 10 'utilities' for $1000. In this case the person is risk averse; he maximizes expected utility but not expected income.

In this context, the maximin principle (as well as multiplication of utilities) in a veil of ignorance makes no sense for the consistency of the utility axioms

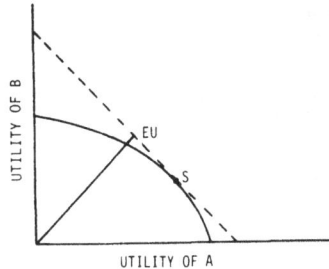

Fig. 13. Maximizing expected utility in a complete veil of ignorance. *S* will be chosen. *EU* is the expected utility of a person.

is violated. If a person were to choose that distribution which maximized the minimum utility, then the undesirability of worse positions should be characterized by the low utility of these positions in the first place.

B. *A Comparison of Maximin 'Income' and Sum of 'Income' Criteria When There is Risk*

In order to make some sense of the maximin principle, under risk, it is useful to discuss not a utility frontier but an income frontier. (Looking at Figure 5 we relabel the axes *A*'s income and *B*'s income, their indifference curves remain the same). Individuals would still be maximizing expected utility. If they were risk neutral (i.e., they were indifferent to a guaranteed million dollars and a fifty-fifty chance of zero or two million dollars), then maximizing expected utility would be equivalent to maximizing expected income (or equivalently the sum of *incomes*). With the axes in income terms, risk neutrality implies that maximizing expected utility creates indifference curves shaped like the sum of utilities curve (when the axes were in utility terms). In contrast, if the person were totally risk averse, maximizing expected utility would create (when the axes were in income terms) an indifference curve looking just like the maximin criterion (when the axes were labelled in utility terms). Clearly, most people are in between perfect risk neutrality and perfect risk aversion.[19]

Viewed from another angle, in a veil of ignorance under risk, constant marginal utility of income means that a person is risk neutral, will maximize expected income and will choose *S* (the point which maximizes the sum of *incomes*). Decreasing marginal utility of income means that the person is risk averse, the more risk averse the closer the indifference curve will approximate the maximin *income* criterion indifference curve and the less it will look like the sum of incomes criterion indifference curve. A fuller treatment of income will take place in Section V.

C. Maximin Criterion With Uncertainty

Under a veil of ignorance, if there is uncertainty (i.e., no probability estimates) regarding the outcome, a maximin strategy is more appropriate than the sum of utilities criterion.[20] Thus the distinction between Rawls and the utilitarian can be seen as one of uncertainty versus risk.

IV. PARTIAL EMPATHY AND A PARTIAL VEIL OF IGNORANCE

A. Partial Empathy

All the justice criteria discussed so far require knowledge of everyone's utility function. This involves enormous informational requirements. Even pareto optimality requires knowledge of everyone's preference ranking (although ordinary markets in the absence of externalities will achieve pareto optimality without anyone knowing anyone else's preference rankings). Furthermore, with the exception of the pareto criterion and the product of utilities (Nash) criterion, all the justice criteria involve either cardinal or ordinal interpersonal comparisons of utility (complete empathy). Since there is no objective way to make interpersonal comparisons of utility, these criteria may never be able to be used as a tool for ethical decision-making.[21] The problems involved with complete empathy have led to measures which only involve partial empathy and thus do not involve interpersonal comparisons of utility.[22] Unfortunately, as will be seen, these methods have serious problems of their own.

For simplicity, assume that a person's utility depends only upon the goods and services he receives. Partial empathy means that the person can place himself in the other person's 'shoes' but he still retains his own utility function. Thus we have a partial veil of ignorance. Under a social contract theory the person knows his utility function but not which objective position he will be in after the contract is made.

With partial empathy both *A* and *B* provide estimates of the utility tradeoff frontier. Person *A* measures everything in terms of *A*'s utility function. Thus, if *A* receives shoes and *B* receives a coat, *A* first measures the amount of

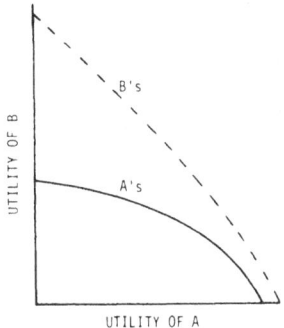

Fig. 14. Tradeoff frontier measured in terms of each person's own utility function. ——
Utility tradeoff frontier in terms of A's utility function. – – – Utility tradeoff frontier
in terms of B's utility function.

utility A has in this situation. A then measures B's utility as if B had the same kind of utility function as A; that is, A measures how much utility A would have if A received a coat and B received shoes; he then attributes this utility to B. B does likewise. Thus the utility tradeoff frontier is different from that encountered in the first three parts as these two curves are each based on only one person's utility function (see Figure 14). Having determined the utility tradeoff frontier, one can apply the same set of justice indifference curves as used in parts 1 and 2.

B. Problems With Partial Empathy

Notice that partial empathy does not involve interpersonal comparisons of utility. Unfortunately, this also may result in basic disagreement between A and B even if they use the same justice criterion. In Figure 15, both A and B use the sum of utilities approach. In this example, in terms of A's utility function, giving everything to B maximizes the sum of utilities. In terms of B's utility function, giving nearly everything to A maximizes the sum of utilities. This is a likely possibility, for our reasoning in using partial empathy in the first place was that utility functions are different from one person to another (and it was therefore hard to be fully empathetic).

With different just allocations suggested by different people, we are in a bind. Social choice theory shows that in general, there are no reasonable methods of aggregating preferences (just or otherwise).[23] Furthermore, a just solution need not even be pareto optimal as the following example will

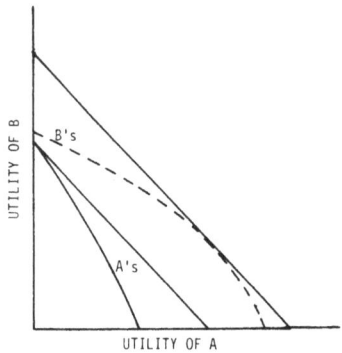

Fig. 15. Maximizing expected utility in a partial veil of ignorance.

show: Let A's utility ranking of fruit be: apples, oranges, pears (in order of preference). Let B's utility ranking be: pears, oranges, apples. Let the set of possible allocations be: 1 apple to A and 1 pear to B; 1 pear to A and 1 apple to B; or 1 orange to each. By the maximin criterion with partial empathy (M^{**}) both would choose 1 orange to each, although both of them would be better off if A received one apple and B one pear.[24]

Lack of pareto optimality can be illustrated in another context. Making the common economic assumption that there is diminishing marginal rate of transformation in the production of goods and services (the *production* possibility curve is bowed out), that marginal utility of goods is positive with decreasing marginal utility and with diminishing marginal rate of substitution (*utility* indifference curves are bowed in), and that feasible allocations are independent of the labelling of the participants (i.e., if one feasible allocation is for A to receive shoes and B to receive a raincoat, it is also feasible for B to receive shoes and A to receive a raincoat), then the utility tradeoff curves are all symmetric around the 45° line. Therefore, any symmetric justice function (all that we have explicitly considered so far) will provide that all goods and services are divided equally. This equal division of goods and services will not in general result in equality of utility or pareto optimality (as some participants may want to trade with other participants for mutual advantage).

C. Minimal Envy Criterion

The criterion of minimal envy also uses partial empathy.[25] With this criterion the person compares his utility with the utility that he would derive if he were

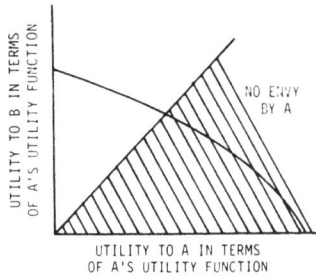

Fig. 16. No envy indifference 'curve'. *A* is not envious of the shaded portion. The utility tradeoff frontier is measured in terms of *A*'s utility function.

in someone else's position. Utility tradeoff frontiers in terms of *A*'s or *B*'s utility, as drawn in Figures 14 and 15, are employed. If the person derives at least as much utility from being in his own position as being in anyone else's position, then the allocation has no envy for the player. In this case the veil of ignorance is removed entirely for the person knows his own utility and the position of all the others. Each participant employs a fat indifference 'curve'. For *A*, any point below the 45° line involves no envy (see Figure 16); similarly, for *B*, any point above the 45° line involves no envy.

Because the axes are labelled in terms of utility rather than allocation to individuals, it is possible for a particular allocation (say shoes to *A* and coat to *B*) to be below the 45° line when measured by *A* who prefers shoes to coats and to be above the 45° line when measured by *B* who prefers coats to shoes. It is thus useful to consider a diagram which involves both allocation and utility functions. We make use of the Edgeworth-Bowley box (Figure 17). This will enable us to discover whether there exists allocations such that no player is envious.

Allocations of goods *X* and *Y* are measures along the horizontal and vertical axes respectively. Allocations to *A* are measured from the lower left hand corner, while allocations to *B* are measured from the upper right hand corner. Thus, position *Z* allocates x_A, y_A to *A* and $X - x_A$, $Y - y_A$ to *B*. If one makes the ordinary economic assumptions (which are not necessary for the analysis) concerning utility functions, then *A*'s and *B*'s indifference curves will be bowed in towards their origin as drawn.

In Figure 17b, Z^* is at $X/2$, $Y/2$. At Z^* neither *A* nor *B* are envious of each other. *A* is not envious of *B* at any point to the right of *A*'s indifference

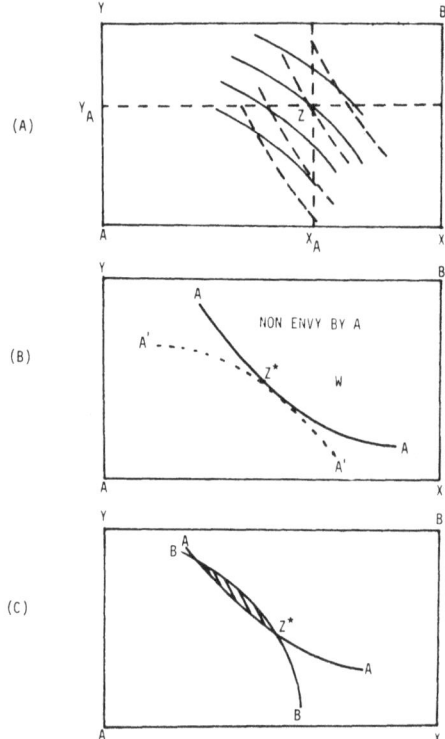

Fig. 17. *A. Indifference curves:* – – – *A*'s indifference curves. Higher utility: upwards and to the right. —— *B*'s indifference curves. Higher utility: lower and to the left. *B. Non-envy:* Note that non-eny area will include part of the space between *A-A* and *A'-A'*. *C. Non-envious allocations Pareto superior to Z*.*

curve (*A*) through *Z**. Any point (*W*) to the right of indifference curve *A* is preferred by *A* to *Z**. If *A* were in *B*'s shoes with *A*'s indifference curve he would have indifference curve A^1. Thus, *A* cannot view *B* as being better off when point *W* is chosen.

In Figure 17c, both *A* and *B*'s indifference curves are drawn through *Z**. The area of overlap is an area of non-envy by both *A* and *B* and of higher utility to *A* and *B* than position *Z**. So, while position *Z** is a point of zero envy, it is not pareto optimal. However, if free trade is allowed, then the final allocation will not only be free of envy but pareto optimal as well. If a person trades, it is because what he gets is worth more than he gives up. Since neither are envious to begin with, and any trade will make *A* better off and *B* worse off from A's point of view and A worse off and B better off from B's point of view, neither will be envious after a trade. If there are third parties, they will not be envious, because they could have outbid but did not, thereby showing that they preferred not to trade than to trade.

With freely transferable and divisible goods, a pareto optimal and non-envious position can be established. However, without these characteristics, one or the other may not be achieved. Two examples will illustrate. If there is only one shirt available, and both A and B desire the shirt (in one piece), no shirt to either is a position of non-envy but violates pareto optimality (if envy does not enter into the utility function but is only a criterion for social choice) as A would be made better off by having the shirt and B would not be worse off as he would not have the shirt in either event.

If A is good looking but would rather be athletic while B is athletic but would rather be good looking, the situation is pareto optimal (as there are no viable alternatives) but one of complete envy.[26]

If everyone had identical utility functions then equal utility would imply 0 envy. If all goods were freely transferable and divisible and everyone had identical concave utility functions, then there would exist a point of 0 envy which was also a point of equal utility (E) and a maximin point (M). If the basis of people's envy was others' happiness rather than their objective circumstances, then equal utility would be the only point of zero envy.

V. INCOME EQUALITY

In this section we consider a number of justifications made for and against equality of income. As will be seen, both sides of the argument rely on very special cases.

It is first assumed that utility is solely a function of income.[27]

PROPOSITION 4. If the income tradeoff frontier is concave and symmetric around the equality line and no one is a risk preferrer, then perfect equality satisfies the maximin criterion with partial empathy and the sum of utilities criterion with partial empathy (see Figure 18).

The income tradeoff curve is symmetric; partial empathy maintains this symmetry in the utility tradeoff curve as the person values a given amount of income to A as having the same value as the same amount to B by partial

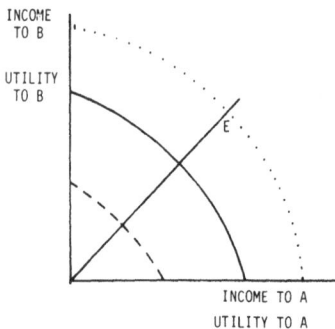

Fig. 18. Equivalent results. If the *income* tradeoff frontier is concave and symmetric and the utility functions are concave. then $E = M = S$. The two graphs (income, utility) are superimposed. —— Income tradeoff frontier. ···· Utility tradeoff frontier measured in terms of A's utility function. – – – Utility tradeoff fronter measured in terms of B's utility function.

empathy. The income tradeoff curve is concave; lack of risk preference maintains concavity in the utility tradeoff curve. By similar logic to that used for Proposition 2, equality is a maximin point and a maximum sum of utilities point for each person.

PROPOSITION 5. If all the utility functions are identical and the conditions of Proposition 4 hold, then equal income satisfies the maximin and sum of utilities criteria (with complete empathy). The proof is obvious for with identical utility functions complete empathy is equivalent to partial empathy.

Those who have argued for inequality have used four types of arguments:

(1) that the income or utility curves are of a different shape;

(2) that there is disutility in producing income;

(3) people deserve the fruits of their contribution;

(4) there should be equality of opportunity but not necessarily of result.

In general the presence of non convexities in either the distribution of income or the utility of income will tend to create inequalities. In fact, if the total income is fixed and people are risk preferrers (or equivalently have increasing marginal utility of income) in a veil of partial ignorance, they will choose an unequal distribution of income with all the income going to one person. As another example, equal income may cause total income to be less. While this situation may still be pareto optimal, people might prefer the unequal distribution as their expected income (in a partial veil of ignorance) would be higher and this might result in a higher expected utility (in the partial veil of ignorance). While this position is often taken by free market proponents, it is not true that the market naturally leads to maximization of the sum of utilities. All that a market can do (when working properly) is achieve pareto optimality.

The second and third arguments are much more common; however, at present the underlying logic behind these arguments is missing. For example, a popular argument for inequality of income is that people who are taxed have less utility than those who are subsidized as the former must undertake the unpleasant task of working harder. Equality of after tax income thus denies the high income earners compensation for the disutility of extra work and consequently they have less after tax equality (measured in utility terms). This only makes sense if the taxation was instituted after the person earned his income.[28] However, if the person knew he was going to be taxed, he clearly has greater utility working hard and being taxed than not working hard and being subsidized, for he freely chose the former (and by the same logic the person who was subsidized preferred not to work as he freely chose the latter). So this argument for inequality of income falls apart.

If people have the same utility functions for income and leisure but different productivities, then equal utilities (made as large as possible) implies that the more productive work more and the less productive work less and the more productive receive more after tax income. The tax is a lump sum tax so as not to distort incentives.

If people are equally productive, but differ in their taste for work such that their utility from leisure is the same but some get more utility from goods and services than others (this means a cardinal interpersonal comparison of utility), then those with a taste for work should work more, be taxed (with a lump sum tax) and their after tax income should be less than those who do not have a taste for work. On the other hand, if people have the same taste for work but some have greater taste for leisure, those who have a taste for leisure should work less and receive less money income.

The third main argument is on marginal productivity grounds. The person deserves to receive his contribution to total income. Each according to his contribution is not a very clear cut guide for it depends on the initial endowments and states of the world and since justice need not treat the status quo as the desirable initial condition, marginal productivity ignores the major issue. For example, even with identical utility functions we can expect the children of wealthy parents to be more productive (as they have more capital) than children of poor parents. Thus an initial change in the distribution of wealth would result in a change in people's contribution to income. Thus marginal productivity cannot be the justification as it in turn depends on the initial endowments.

So what should determine the initial endowments? In many ways we are back to the first three parts. For example, we may choose the initial endowments so that the end result is equality of utilities, maximin utility, equality of income, etc. One possibility, not already discussed, is equal opportunity. It is to this issue we now turn.

There are two levels of equality of opportunity. In the first type, a specific type of opportunity is open to all – e.g., free university education. This is not tailored to the particular endowments of individuals (e.g., differences in skills, parents' income). The second type equalizes all objective endowments. For example, the person with low intelligence is compensated with extra teaching or physical capital. If people have identical utility functions, the second type of equal opportunity will lead to equal income and utility.

University of California, Santa Cruz

NOTES

* This paper is a revised version of a lecture delivered at the University of Chicago (1976) and at the Public Choice Meetings in New Orleans (1978).
[1] Some of the discussion can also be related in terms of measurables such as income instead of utility. Most of the analysis is in the context of a two person model so that we can use a simple diagram.
[2] Thus the line defined by the set of pareto optimal points slopes downwards from left to right and is not concerned with the curvature.
[3] The maximin criterion was formulated by Rawls (1971) within a different context (a veil of ignorance). See Hammond (1976a), Sen (1978b, 1977b), Deschamps and Gevers (1978), and Strasnick (forthcoming) who compare Rawls' approach to the utilitarian approach [Bentham (1838) and Mill (1876)]. For a critical discussion of Rawls from an 'economic' viewpoint see Musgrave (1974), Harsanyi (1975b) and Rae (1975). Important surveys on justice include Sen (1970, 1977b) and D'Aspremont and Gevers (1977).
[4] See Harsanyi (1955), Sen (1977b), Maskin (1978), Deschamps and Gevers (1978), and Roberts (forthcoming) who provide axiomatic justification for utilitarianism.

[5] Cardinality allows summation. For example, money has cardinality: $60 + $30 = $90. With (full) cardinal interpersonal comparisons of utility, a different origin or unit of measurement for A or B's utility function will yield different results.

[6] Since both criteria weigh utility neither method disallows 'unjust' acts from occurring. E.g., the benefit to the Nazi may be greater than the disbenefit to the Jew (sum of utilities) while the unhappiness of the Nazi if not allowed to kill may be greater than the unhappiness of the Jew in being killed (maximin). This is why Rawls requires in 'Justice as fairness' (1958) men of equal stature.

Rawls infortunately equates the word 'efficiency' with maximization of the sum of utilities. However, this is not how modern economists use the word, as efficiency is another word for pareto optimality. If people truly are in a veil of ignorance then the sum of utilities is *ex ante* pareto optimal (see Section III). For a valuable discussion of the maximin versus utilitarian approaches see Blackorby and Donaldson (1977) who use related diagrams in their presentation.

[7] In Section III they will be shown to be the limiting cases in attitudes towards risk.

[8] For discussions of the Gini coefficient and social welfare, see Blackorby and Donaldson (1978), Sen (1976a) and Kolm (1976a, b). The relationship between the Gini coefficient and the maximin criterion only hold for two people.

[9] The percent egalitarian criterion measures inequality as the percent difference in utility (instead of absolute difference). In this case the indifference curves are rays (straight lines) starting at the origin and fanning out. The highest indifference curve is again the 45° line from the origin. See D. Wittman (1974) for theories of justice based on pure distribution.

[10] This may be due to economies of scale, externalities, different most preferred outcomes, etc.

[11] If the pareto frontier is a 45° line to the axes, then many points satisfy the sum of utilities criterion but only one satisfies the maximin criterion.

[12] This proof only applies for 2 people.

[13] The concept of a social contract is very important in political philosophy (e.g., Locke (1960) and Rousseau (1954); a more recent example is Buchanan and Tullock (1962)). Needless to say we will not consider the political aspects here. Recent works using the concept of the social contract in arriving at just outcomes are Mueller (1975); Mueller *et al.* (1976); Pazner and Schmeidler (1977).

[14] For a discussion of this problem see Harsanyi (1975). Also see Pazner and Schmeidler (1976).

[15] That is, we do not want (full) cardinality but only the 'partial cardinality' implied by the von Newmann-Morgenstern utility function.

[16] See Nash (1950) who did not see this method specifically as a justice criterion. See Brock (1979) for a recent treatment. The Shapley value (Shapley (1953)) with transferable utility is a special case of the Nash solution as the additional utility over the status quo will be divided equally. This maximizes the product of the additional utilities.

[17] Analogous issues are found in game theory. In the solution to the cooperative game, is the status quo the maximin strategies (Shapley, 1950) or the threat strategies (Nash, 1953, not to be confused with Nash, 1950, where the determination of the status quo is unspecified)?

[18] See Harsanyi (1953, 1955a, 1975b) who argues in favor of the expected utility hypothesis for justice criteria and Diamond (1967), Rawls (1971), Sen (1970, 1973, 1977a) who argue against.

[19] By symmetry the possibility of a risk preferrer should be considered. Being the complete optimist he would only weigh the maximal utility. The indifference curves would be shaped as follows:

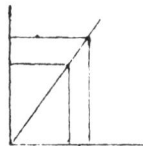

[20] The maximin criterion is appropriate for a zero sum game. No strategy is correct for a non-zero sum game.

[21] Within orders of magnitude one can establish some consensus on interpersonal rankings. For example, with some search one could find two people with similar objective circumstances and yet almost everyone would agree that one of the two people was happier (had greater utility) than the other who was constantly angry or crying. As another example, we may all agree that A, who likes scuba diving, will get greater utility by being allowed to scuba dive for the rest of his life in exchange for person B playing one less game of pinochle even though we are deathly afraid of scuba diving and would get no pleasure from the activity.

[22] See Pattanaik (1968) for a comparison between partial and full empathy.

[23] Although social choice mechanisms for aggregating just preferences may not need to satisfy the same criteria. For example, as will be shown, M^{**} need not even be pareto optimal.

[24] Some might claim that this result is unjust and that justice requires taking into account the person's preferences for what he receives. That is, interpersonal comparisons are required for justice.

[25] See Varian (1974, 1976a, b), Feldman and Kirman (1974), Schmeidler and Vind (1972) and Daniel (1975). They are generally concerned with fair allocations which have zero envy and in addition are pareto optimal.

[26] Daniel (1975) has called this a just position as the same number of people envy person A as A envies.

[27] See Sneed (1975) and Musgrave (1974) for a discussion of leisure.

[28] This error is commonly found in anarchist theories — that the state of nature is without government and that government changes the rules after the fact. See Friedmann (1962.)

BIBLIOGRAPHY

Arrow, K. J.: 1977, 'Extended sympathy and the possibility of social choice', American Economic Review 67, pp. 219–225.

Atkinson, A. B.: 1970, 'On the measurement of inequality', Journal of Economic Theory 2, pp. 244–263.

Barry, B.: 1973, The Liberal Theory of Justice (Clarendon Press, Oxford).

Barry, B. and D. W. Rae: 1975, 'Political evaluation', in N. Polsby and F. Greenstein (eds.), The Handbook of Political Science, Vol. 1 (Addison Wesley).

Becker, E.: 1975, 'Justice, utility and interpersonal comparisons', Theory and Decision 6, pp. 471–484.

Bentham, J.: 1838–43b, 'Introduction to the principles of morals and legislation', Works, Vol. 1.

Blackorby, C.: 1975, 'Degrees of cardinality and aggregate partial ordering', Econometrica 43, pp. 845–852.

Blackorby, C. and D. Donaldson: 1977, 'Utility vs. equity; Some plausible quasi-orderings', Journal of Public Economics 7, pp. 365–381.

Blackorby, C. and D. Donaldson: 1978, 'Measures of equality and their meaning in terms of social welfare', Journal of Economic Theory 18, pp. 59–80.

Brock, H. W.: 1979, 'A game theoretic account of social justice', appearing in the present double issue of Theory and Decision.

Buchanan, J. M. and G. Tullock: 1962, The Calculus of Consent (The University of Michigan Press, Ann Arbor).

Daniel, T. E.: 1975, 'A revised concept of distributional equity', Journal of Economic Theory 71 (1), pp. 94–109.

D'Aspremont, C. and L. Gevers: 1977, 'Equity and the informational basis of collective choice', Reveiw of Economic Studies 46, pp. 199–210.

Deschamps, R. and L. Gevers: (forthcoming), 'Leximin and utilitarian rules: A joint characterization', Journal of Economic Theory.

Diamond, P.: 1967, 'Cardinal welfare, individualistic ethics and interpersonal comparisons of utility: A comment', Journal of Political Economy 61, pp. 765–766.

Feldman, A. and A. Kirman: 1974, 'Fairness and envy', American Economic Review 64, pp. 995–1005.

Fine, B.: 1956, 'A note on "Interpersonal comparisons and partial comparability"', Econometrica 43, pp. 169–172.

Fleming, M.: 1973, 'A cardinal concept of welfare', in E. S. Phelps (ed.), Economic Justice (Penguin Books, Harmondsworth), pp. 245–265. First published: Quarterly Journal of Economics, Vol. 66, 1952, October.

Friedman, M.: 1962, Capitalism and Freedom (University of Chicago Press, Chicago).

Hammond, P.: 1977, 'Dual interpersonal comparisons of utility and the welfare economics of income distribution', Journal of Public Economics 7 (1), pp. 51–73.

Hammond, P. J.: 1976a, 'Equity, Arrow's conditions and Rawls' difference principle', Econometrica 44, pp. 793–804.

Hammond, P. J.: 1976b, 'Why ethical measures of inequality need interpersonal comparisons', Theory and Decision 7, pp. 263–274.

Harsanyi, J. C.: 1953, 'Cardinal utility in welfare economics and in the theory of risk-taking', Journal of Political Economy 61, pp. 434–435.

Harsanyi, J. C.: 1955a, 'Cardinal welfare, individualistic ethics and interpersonal comparisons of utility', Journal of Political Economy 63, pp. 309–321.

Harsanyi, J. C.: 1975b, 'Non-linear social welfare functions, or Do welfare economists have a special exemption from Bayesian rationality', Theory and Decision 6, pp.

311–332.

Harsanyi, J. C.: 1975b, 'Can the maximin principle serve as a basis for morality: A critique of John Rawls's theory', American Political Science Reveiw 69, pp. 594–606.

Kolm, S. C.: June 1976, 'Unequal inequalities 1', Journal of Economic Theory 12 (3), pp. 416–442.

Kolm, S. C.: August 1976, 'Unequal inequalities 2', Journal of Economic Theory 13, pp. 82–111.

Locke, J.: 1960, Two Treatises of Government, P. Laslett, Intro. and Notes (New American Library, New York).

Maskin, E.: 1978, 'A theorem on utilitarianism', Review of Economic Studies 45, pp. 93–96.

Mill, J. S.: 1867, Utilitarianism, 3rd edition.

Mueller, D. C.: 1975, 'Uncertainty, information, and redistributive justice', Draft (Cornell University (mimeo)).

Mueller, D., R. Tollison, and T. Willett: 1976, 'The utilitarian contract: A generalization of Rawls' theory of justice', in R. C. Amacher, R. D. Tollison, and T. D. Willett (eds.), The Economic Approach to Public Policy, (Cornell University Press, Ithaca), pp. 313–333.

Musgrave, R.: 1974, 'Maximin, uncertainty and the leisure trade off', Quarterly Journal of Economics 88, pp. 625–632.

Nash, J. F.: 1950, 'The bargaining problem', Econometrica 18, pp. 155–162.

Nash, J. F.: 1953, 'Two-person cooperative games', Econometrica 21, pp. 128–140.

Pattanaik, P. K.: 1973, 'Risk, impersonality and the social welfare function', in E. S. Phelps (ed.), Economic Justice (Penguin Books, Inc., Harmondsworth), pp. 298–318. First published: Journal of Political Economy, December, 1968.

Pazner, E. A.: 1977, 'Pitfalls in the theory of fairness', Journal of Economic Theory 14 (2), pp. 458–466.

Pazner, E. A. and D. Schmeidler: 1976, 'Social contract theory and ordinal distributive equity', Journal of Public Economics 5 (3–4), pp. 261–268.

Phelps, E. S.: 1973, Economic Justice (Penguin, Harmondsworth).

Rae, D.: 1975, 'Maximin justice and an alternative principle of general advantage', American Political Science Review 69 (2), pp. 630–647.

Rawls, J.: 1958, 'Justice as fairness', Philosophical Review 67.

Rawls, J.: 1971, A Theory of Justice (Harvard University Press, Cambridge, Mass., and Clarendon Press, Oxford).

Rawls, J.: 1974, 'Reply to Musgrave', Quarterly Journal of Economics 88, pp. 633–655.

Roberts, K. W.: forthcoming, 'Interpersonal comparability and social choice theory', Review of Economic Studies.

Rousseau, J. J.: 1954, The Social Contract (Henry Regnery Co., Chicago), W. Kendall, trans. and introduction.

Schmeidler, D. and K. Vind: 1972, 'Fair net trades', Econometrica 40, pp. 637–642.

Sen, A. K.: 1970, Collective Choice and Social Welfare (Holden Day, San Francisco, and Oliver & Boyd, Edinburgh).

Sen, A. K.: 1973, On Economic Inequality (Clarendon Press, Oxford, and Norton, New York).

Sen, A. K.: 1976a, 'Poverty: An ordinal approach to measurement', Econometrica 44, pp. 219–231.

Sen, A. K.: 1976b, 'Welfare inequalities and Rawlsian axiomatics', Theory and Decision 7, pp. 243–262; also in R. Butts and J. Hintikka (eds.), Logic, Methodology and Philosophy of Science (Reidel, Dordrecht, 1976).

Sen, A. K.: 1977a, 'Non-linear social welfare functions: A reply to Professor Harsanyi', in R. Butts and J. Hintikka (eds.), Logic, Methodology and Philosophy of Science (Reidel, Dordrecht).

Sen, A. K.: 1977b, 'On weights and measures: Informational constraints in social welfare analysis', Econometrica 45 (7), pp. 1539–1572.

Shapley, L. S.: 1953, 'A value for n-person games', in H. W. Kuhn and A. W. Tucker (eds.), Contributions to the Theory of Games II, Annals of Mathematics Studies 28 (Princeton University Press, Princeton).

Sneed, J. D.: 1975, 'Some consequences of Rawlsean justice in simple economies', Seminar für Philosophie, Logik und Wissensehaftstheorie, Universität München, September 20, 1975.

Strasnick, S.: forthcoming, 'Social choice theory and the derivation of Rawls' difference principle', Journal of Philosophy.

Suppes, P.: 1966, 'Some formal models of grading principles', Synthese 6, pp. 284–306; reprinted in P. Suppes, Studies in the Methodology and Foundations of Science (Reidel, Dordrecht, 1969).

Varian, H.: 1974, 'Equity, envy and efficiency', Journal of Economic Theory 9, pp. 63–91.

Varian, H.: 1975, 'Distributive justice, welfare economics and the theory of fairness', Philosophy and Public Affairs 4, pp. 223–247.

Varian, H. R.: 1976a, 'Two problems in the theory of fairness', Journal of Public Economics 5, pp. 249–260.

Varian, H. R.: 1976b, 'On the history of concepts of fairness', Journal of Economic Theory 13 (3), pp. 487–7.

Wittman, D.: 1974, 'Punishment as retribution', Theory and Decision 4, pp. 209–237.

HORACE W. BROCK

A GAME THEORETIC ACCOUNT
OF SOCIAL JUSTICE

ABSTRACT. The role in ethics of game theory proper (as opposed to decision theory) is discussed via an elucidation of a new theory of justice. The new theory integrates into a coherent whole two fundamental distributive norms: To Each According to his Needs; and to Each According to his Contribution. The theory incorporates a new account of ethics in terms of impartial decision – an account which dispenses with the need for a Veil of Ignorance construct. Also, the new theory does not require the use of interpersonal comparisons of utility at an operational level, even though such comparisons arise at a conceptual level. The reason for this lies in its relationship to game theoretical structures which do not entail interpersonal comparisons. Finally, the theory makes possible a new interpretation of two cooperative game solutions: The Nash solution, and the Generalized Shapley Value.

I. INTRODUCTION

In recent years, decision theory has provided the basis for some searching investigations into the problem of social justice. And yet game theory proper, which is a generalization of decision theory, has not played a correspondingly important role in ethics. The purpose of the present essay is to explore in an informal manner a new, game theoretic account of justice[1]. The interested reader can find a formal mathematical version of this theory elsewhere (Brock, 1978). In discussing this new theory, I intend to place special emphasis on revealing the ways in which game theory can be helpful in a moral philosophical context.

There are three salient features of the proposed theory. First, it integrates into a coherent whole two ethical subtheories. These subtheories reflect a concern with two differing concepts of distributive justice: allocation according to relative need; and allocation according to relative contribution. Moreover, each subtheory is given an unambiguous game theoretic characterization. Second, the theory incorporates a new concept of impartiality which is inspired by and congruent with the concept of relative needs and which dispenses with the concept of rational choice under uncertainty. Third, while

Theory and Decision 11 (1979) 239–265.   0040–5833/79/0113–0239$02.70.

the theory makes an essential use of interpersonal comparisons of utility at a conceptual level, it does not entail such comparisons at an operational level.

In Section II I furnish the reader with an overview of the basic stucture of of the new theory. In Section III, the subtheory of distribution according to relative needs is developed. Here attention is drawn to the mathematical equivalence of my ethical model with the Nash-Harsanyi theory of pure n-person bargaining games, and with the Kaneko-Nakamura version of axiomatic social choice theory. In Section IV I briefly discuss the subtheory of distribution according to relative contribution. And in Section V I present an informal characterization of a two-stage game, a play of which will realize 'full distributive justice' as understood in the new theory.

II. OVERVIEW OF THE THEORY

On a rather broad level, the theory is based on my belief that there exist two quite different yet equally compelling concepts of distributive justice: allocation in accord with relative need (RN); and allocation in accord with relative contribution (RC)[2]. A purpose of the theory is to provide a satisfactory characterization of each of these two norms, and to demonstrate their proper interrelationship. The first of these tasks will be addressed in the next two Sections of this paper. It is the latter task I wish to address in the present section.

For my purposes in the present paper, I shall simply assume that there are two fundamental distributive norms, namely the RN and RC norms just mentioned. Let us introduce alongside of these norms two different kinds of environments: manna and non-manna environments[3]. By a *manna environment* I shall mean an environment in which people do not have differential claims on the social product that are due to differential contributions to production of the product. For example, consider the case of two children who wander into the kitchen and find an apple pie on the kitchen table, – a pie which neither has baked or bought. Then it seems reasonable to view the pie as manna from Heaven as far as the two children are concerned. Accordingly, they (or their guardian) face a manna distribution problem. This situation can easily be distinguished from a non-manna distribution problem which will arise if one or the other – or both – of the children have produced the pie.

Now while it would seem appropriate to invoke some needs-oriented allocative principle in a manna distribution problem, it is not prima facie clear what type of principle should be adduced in a non-manna situation. By demonstrating that a particular *dependence condition* can obtain between the RN norm and the RC norm, I hope to make a case that it is ethically respectable to appeal to a principle of relative contribution (RC) in certain non-manna situations.

II.A. *The Basic Set-Up*

By introducing the basic set-up underlying the theory, it will be possible to establish the mesh between the two fundamental distributive norms. Assume that there exist n people in some hypothetical society. Each of these n people is assumed to have (cardinal) preferences represented by means of a von Neumann-Morgenstern utility function. Whether and in what form interpersonal utility comparisons arise will be made clear in the sequel. Whenever I speak of a game or of an arbitration problem, it will be understood that the decision problem in question is one of complete (if imperfect) information. Moreover it is assumed that any given game has a unique solution, e.g. the 'traced' equilibrium point of the recent Harsanyi-Selten theory (Harsanyi, 1975).

The basic analytical tool in the proposed theory is a *two-stage decision problem*, denoted by G^*. For the moment, I shall discuss G^* in a very informal manner. This discussion will be supplemented by Section V where G^* is characterized as a particular two-stage game the prizes of whose first stage are a set of second-stage games. In Stage I of the decision problem, a constitution must be chosen. This constitutional choice decision problem is denoted G^c. The *choice set* of G_c is simply the set $P(C)$ of all probability mixtures of the alternative possible constitutions. The set of alternative possible constitution is denoted C. Just below I shall show how game theory can be used to characterize C in an unambiguous way.

Now choice of some constitution $c \in C$ is thought to induce an associated second-stage decision problem G_c. The basic idea here is that a constitution serves partially to define the decision problem with which we are confronted in daily life. The choice set of the Stage II decision problem can thus be thought of as the set of all possible constitutionally admissible 'life-plans' — to use a Rawlsian phrase — amongst which the citizens can choose. This set of admissible life plans will depend upon the constitution c chosen in the Stage I problem G^c.

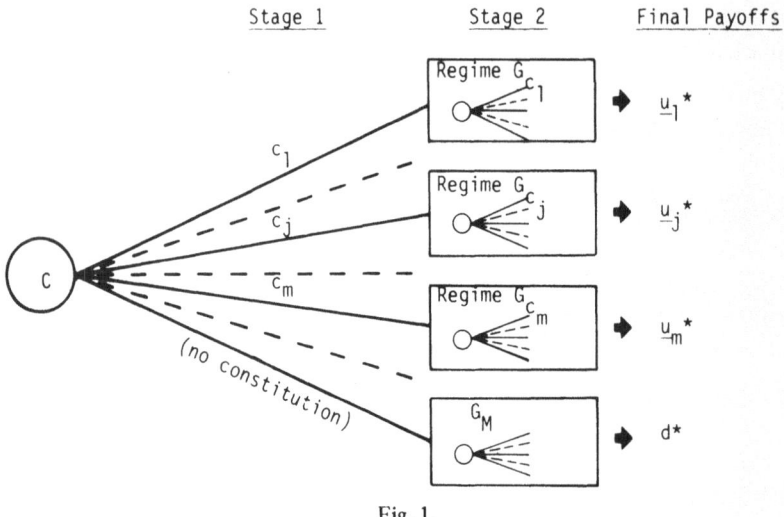

Fig. 1.

The payoffs in G^c and G_c are interdependent in the following sense. The payoff n-tuple generated in G^c by choice of some constitution $c \in C$ is assumed to be the payoff n-tuple awarded the citizens by participation in the c-induced decision problem G_c played in Stage II of the two-stage problem G^*. This interdependence is schematized in Figure 1 where such notation as is introduced should be self-explanatory.

A question arises as to the circumstances under which the Stage I decision problem G^c is meaningful. Why would the n citizens ever consider participating in a constitutional choice problem? I shall introduce the familiar assumption in this context that *all n* citizens can improve their prospects by agreeing to participate in G^c. [Because of the utility gains which can be sponsored by side-payments in most real-world situations, I do not regard this assumption to be very severe.]

More specifically, if they do not participate in G^c, of if they fail to reach agreement in G^c on some constitution, they they will be assumption participate in a *default decision problem G_M* whose payoff n-tuple is dominated by the payoff n-tuple of at least one (and presumably many) constitutional regimes G_c. In short, the citizens have an incentive to reach an agreement in G^c. All will gain by doing so. The payoff from G_M will be denoted by d^*, and can be thought of as the zero point of the constitutional decision problem G^c.[4]

II.B. *The Role of the Two Distributive Norms*

The Stage I decision problem G^c can be interpreted as a manna distribution problem, whereas the Stage II decision problem can be interpreted as a non-manna problem. Participation in G^c amounts to an act of constitutional deliberation. If, as seems reasonable, we assume that all n citizens are 'equal' in these deliberations in the sense of being guided by the same principles of rational choice or ethics, then no one can really be said to contribute more than anyone else. Participation in G^c is in effect a *luxury* which no one has earned or brought about.

But the situation is clearly different in the Stage II problem G_c. Here different people will adopt different life plans, will utilize their differing talents differently, and will generally contribute differentially to such social product as is produced.

My *fundamental proposition* is that (i) the decision problem G^c should be solved (i.e., played/arbitrated) on the basis of a theory of relative need; and (ii) the Stage II problem G_c should be solved on the basis of a theory of relative contribution.

The intuitive justification for (i) is straightforward. Any satisfactory theory of manna distribution will at some level or other be based upon the concept of relative needs.[5] The intuitive justification for (ii) is trickier. Many philosophers feel that the contribution principle has a lesser status than the needs principle. Rawls (1971, pp. 305–309) would seem to share this view, although he never clarifies the relationship between the two norms. My view is that both norms are fundamentally important. However, in order for the contribution principle to be applicable in a given decision problem, the overall environment in the decision problem must be ethically respectable. Specifically, in order that the contribution principle apply in G_c, the constitution c that induces G_c must have been chosen in the prior problem G_c on the basis of relative needs. In short, both norms are indispensable; but there is a specific *dependency condition* which relates the two norms in a critically important way.

A simple example might help at this point. Suppose that it is generally agreed that the 'basic institutions' of our society are just. E.g. they were selected on the basis of some impartial and needs-respecting process. Under these circumstances, how do we react if a very able, hard-working and *contributory* junior faculty member is refused tenure wherease a less talented and contributory person receives tenure? Surely our sense of justice is offended. Specifically, our sense of 'to each according to his contribution' is offended. The question of the relative needs of the two teachers is not likely to arise at all! Yet despite this reality, most contemporary theories eschew the norm of relative contribution.

FUNDAMENTAL DEFINITION. I shall say the *Full Distributive Justice* obtains iff G^c is solved in accord with the needs criterion, and if G_c is solved in accord with the contribution criterion.

Two brief comments are in order at this point. First, I have not said what I *mean* by allocation according to 'needs' and 'contribution'. This will be dealt with in Section III and IV. Second, I have not made clear what I mean by a decision problem which must be 'solved'. Precisely what kind of a procedure can be used to solve the two-stage decision problem G^*? I shall deal with this matter in Section V. In the remainder of the present section, I wish to discuss the role of the default game G_M in the proposed theory of justice.

II.C. *The Zero Point and the Nature of the Choice Set C*

The decision problem G_M is important in the theory for two different reasons. First, the payoff n-tuple from this problem will affect which constitution is selected in the Stage I problem G^c. The reason for this is simply that this payoff serves as the zero point (status quo point) in the constitutional choice problem G^c, *and* the welfare function to be used in solving G^c is sensitive to the zero point. I do not regard this last consideration as problematic. First, any coherent theory of allocation according to relative need will necessarily depend upon the zero point, as I attempt to demonstrate in Section III below. Second, I believe that the proper domain for a theory of distributive justice is the set of *gains* which accrue from adoption of a set of basic institutions (e.g. Rawls, 1971, p. 7).

The second reason why the default game G_M is important in the theory is that it is used to provide a formal characterization of the choice set C of the constitutional decision problem G^c. Hitherto, it has never been quite clear what a 'constitution' or set of 'basic institutions' is. In the new theory, game theory is used to provide an unambiguous description of C. Let me take this opportunity to sketch how this works.

Rawls (1971, p. 55) defines a constitution as

> a public system of rules which define offices
> and positions with their rights and duties . . .
> These rules specify certain forms of
> actions as permissible, others as forbidden,
> and they provide for certain penalties.

This intuitively appealing definition can be modeled game theoretically in the following way. Recall that G_M is the game which will be played if no constitution is adopted in the decision problem G^c. In G_M it is reasonable to suppose that 'anything goes'. Specifically, the strategy set of G_M is *maximal* in the sense that it encompasses all and any physically feasible behavior. To state this is to make a descriptive statement about what is physically *possible* in G_M, not a normative statement about what it would be individually or jointly *rational* to *do* in G_M. Let us couple this observation with an argument I have developed elsewhere (Brock, 1978). The gist of this argument is that it is possible to identify the set of all possible *restrictions* on physically possible behavior with the various components of Rawls' definition of a constitution, to wit (i) prescriptions of certain actions; (ii) proscriptions of certain actions; and (iii) adoption of specific penalty and reward systems. What we end up with is the realization that the set C of constitutions can be viewed as the set of all possible *strategy restrictions* of G_M. Choice of a given constitution c will induce a strategy restricted version of G_M, namely the Stage II regime G_c.

III. ALLOCATION ACCORDING TO RELATIVE NEEDS

The concept of distribution according to relative needs is as ubiquitous as it is ambiguous in ethical theory. Rawlsians, utilitarians and Marxists attempt to show that their theories are consistent with some needs principle. The reason for this surely lies in the fact that the concept of needs is the most familiar and appealing of ethically respectable distributive principles. And yet in my own opinion, no particularly compelling or unambiguous theory of needs has been set forth. In Part A of this Section I shall sketch what I believe to be a satisfactory needs theory. Then in Part B this theory will be contrasted with the treatment of needs in the utilitarian and Rawlsian theories.

The role of the needs theory in my overall theory of justice is to provide the basis for a solution to the constitutional choice problem G^c. Recall that

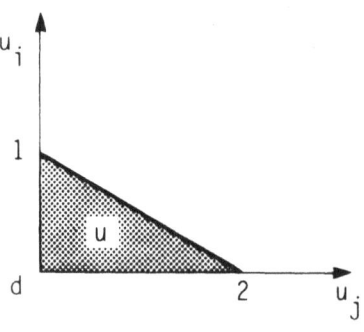

Fig. 2.

this problem is being interpreted as one of manna distribution and therefore calls for a needs-oriented solution concept. In the present Section, I shall suppress any further reference to G^c and shall motivate the needs subtheory by means of some very simple examples.

III.A. *A Theory of Distribution According to Relative Needs*

Consider a simple two-person pie division problem. The *prospect space* consisting of all feasible utility *n*-vectors (with $n = 2$ now) for such a problem is sketched in Figure 2. It is assumed that the two utility functions have been *interpersonally calibrated* with respect to both the *origin* and the *unit* of the utility functions. Later it will become clear in what sense – and why – this assumption can be discarded, at least at an operational level. The origin of the prospect space U will be denoted by the vector d, and is assumed to correspond to the 'no pie' outcome. Note that the Pareto boundary in Figure 2 is *flat*. I have intentionally started off with this special case.[6] When there is a flat boundary, there is an unambiguous measure of relative needs. This is the slope of the Pareto frontier. In the present example, player j gets 2 units of interpersonally calibrated utility for every unit that i gets. I.e., j is twice as needy. Of course, in asserting this, we are in effect equating 'relative needs' with 'relative intensity of desire'. This seems unobjectionable since our concern is with relative – not absolute – needs.

In order to determine an allocation of pie that corresponds to the relative

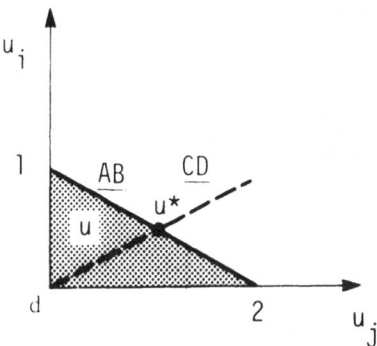

Fig. 3.   The equation of *CD* is (1) below; the equation of *AB* is $\Sigma a_k u_k$ = c (where c is a hyperplane constant).

needs of *i* and *j*, let me introduce my fundamental norm of equitable manna distribution according to relative needs:

POSTULATE A. *The Proportional Priority of Preference Postulate (PPPP)*: Pie should be distributed such that the utility gains to the recipients are proportional to their relative needs. Formally

(1)
$$\frac{u_i - d_i}{u_j - d_j} = \frac{a_j}{a_i}$$

where the variables a_i, a_j correspond to the slope of the Pareto frontier. In the present case, we require that $a_j/a_i = 1/2$.

The PPPP in and of itself is not sufficient to determine a unique solution to the problem. For if the reader will consult Figure 3 he will observe that any and every point on the line *CD* satisfies (1). Therefore we supplement PPPP by a straightforward

POSTULATE B. *Efficiency*: The pie must be distributed in such a way that an efficient (Pareto optimal) outcome is achieved. Formally we require

(2)
$$\underset{U}{\text{MAX}} \sum_k a_k u_k \qquad k = i, j$$

Note that this is not the ordinary efficiency condition. For the a_k here are not variables, but are the given slope values, a_i, a_j. Choice of any other values for these weights would clearly render (2) incompatible with (1). Taken together, (1) and (2) clearly define a unique solution to the distribution problem, namely a utility vector u^* lying midway along the Pareto frontier. At this point, i receives a utility gain $1/2$ as great as j receives, which seems reasonable since i's need is half as great as j's. Geometrically, u^* is the point of intersection of the two lines AB and CD which have the property that their slopes are equal in magnitude but opposite in sign.

Three comments are in order. First, my PPPP makes clear that it is *utility* and not pie that is fundamental in the theory. Pie is only important because it generates utility, i.e. fulfills human needs. And to the extent that 'something' must be mathematically allocated in proportion to needs, it is utility, not pie.

Second, the above formulation falls somewhere between the Rawlsian and the utilitarian formulations. This can be seen from two vantage points. To use S. Strasnick's terminology, in the case of utilitarianism, no one has *Priority of Preference*. Everyone's wants enter on an equal footing. In Rawlsian MAX MIN type theories, the worst-off persons's needs receive absolute priority. In the proposed theory, priority is distributed in proportion to relative needs. This comparison on the basis of preference priority is mirrored by differences in the *weights* which characterize the welfare functions of the three theories. Note in particular that whereas the weights in the utilitarian and Rawlsian welfare functions are *constant*, the weights a_i and a_j appearing in (2) above will *vary* according to the distribution of needs in a given problem.

Third, the proposed theory embodies a form of impartiality which is significantly different from that of rational choice under uncertainty (e.g., Veil of Ignorance impartiality). At this point in the paper I shall simply introduce the new concept of impartiality. In Section III.B. I will contrast it with the other concept, and defend it. Let me suggest that a decision is *impartial in an ethical sense* if the *only information* used in arriving at the decision is *ethically significant information*. In a manna distribution problem, this translates into the condition that the decision must be made *solely* on the basis of relative need. E.g., an ethical decision maker will not be influenced by how physically attractive person i is relative to person j, or by his personal preferences for i and i's personalities.[7]

An ostensible limitation of the foregoing account is that it assumes a 'flat' Pareto frontier in order that a well-defined measure of relative need exist. In

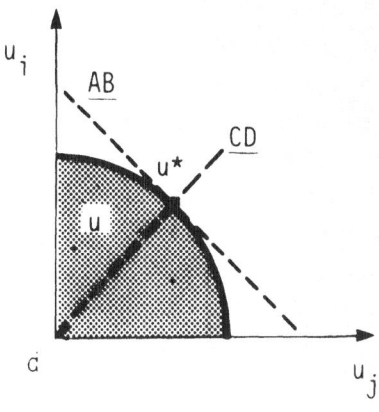

Fig. 4.

general, the Pareto frontier will not be flat, but will be strictly (or at last piecewise strictly) convex. Fortunately, the proposed theory goes through in this more general situation. The reader can convince himself that when the boundary of U is strictly convex, there is once again a unique solution to (1) and (2). See Figure 4 for a schematization. There are two salient differences now, however. First, it is no longer meaningful to suppose that the efficiency condition (2) 'supplements' (1). Rather both conditions enter on equal footing, and the values of the variables a_i, a_j as well as of the payoffs $(u_i - d_i)$, $(u_j - d_j)$ are determined *simultaneously* when (1) and (2) are solved. This is to be contrasted to the earlier case where the values of a_i and a_j were known from the outset. A second difference is that whereas we originally had a *global* measure of relative need, we now have only a *local* measure, namely the slope of the line AB tangent to the solution u^*. However, we can justify an identification of this local measure with a global measure on the following grounds. For any and every strictly (or piecewise) convex problem U there will be a unique *flat problem* $\bar{U}$ constructed as follows. Simply replace the original problem U with the problem $\bar{U}$ whose prospect space is the set of utility vectors bounded by the line AB which is tangent to the solution u^* to the original problem. This flat problem $\bar{U}$ will clearly have the same solution u^* as the original convex problem, so that the two problems can be deemed *solution equivalent*. And the measure of relative needs in both problems will clearly be the same, namely the slope of the line AB. We can therefore let this slope serve as a global measure of relative need in both U and $\bar{U}$.

Note the important role of the disagreement payoff d in the proposed theory. The solution a^* and u^* to (1) and (2) clearly will depend upon the parameter d. To point this out does not prejudge whether there exists a meaningful zero point in a given problem. Often there will not be one. However, what this does suggest is that *if* we want an ethical theory that is firmly based upon an unambiguous account of relative needs, then a zero point will necessarily enter into the matter at some point.[8]

It is now time to state one of the more interesting facts about the theory of manna equity that has been sketched. The theory turns out to be intimately related both to the axiomatic theory of bargaining, and to the axiomatic theory of Arrowian social choice. Loosely, we have a

FUNDAMENTAL CORRESPONDENCE THEOREM. The ethical theory of manna distribution sketched above is fundamentally different from the Nash-Harsanyi theory of bargaining, and from the Kaneko-Nakamura account of social choice theory in a cardinal utility context. Nonetheless, the three theories are *solution equivalent* in the sense that they will always imply the same physical solution to a given problem.

No formal proof of this result will be given here (see Brock, 1978). However I shall provide an informal motivation and justification for the result. The manna theory proposed above is properly speaking a contribution to ethics and welfare economics – not to means-ends rational choice theory as we ordinarily think of the latter. Harsanyi has insistently and rightly pointed out the need to separate these two sets of concerns (e.g., Harsanyi, 1961). Now the purpose of the proposed theory is to characterize distributional equity in terms of the players' relative needs. Accordingly it was both necessary and meaningful to introduce interpersonal comparisons of utility. Or to state the matter differently, it would have made no sense to have introduced a game theoretic (or social choice theoretic) postulate of invariance under separate transformations of the players' utility functions. For the relative needs of a given pair of players in a given situation can only be represented by a *single needs ratio* which will clearly depend upon the calibration of the utilities.

The situation is completely different when we turn to bargaining theory. Here interpersonal comparisons are not necessary since (i) the focus is on the question of rational strategic behavior – not on the fair distribution of utility; and (ii) it is possible to characterize rational behavior in terms of certain dimensionless numbers which Harsanyi (1977a) is called the players' *risk limits*, and which are invariant under separate transformations of the players' utilities.

And yet it happens that the solution prescribed by the Nash-Harsanyi theory (i.e., maximize the product of the utility gains) can be restated in the form of equilibrium conditions which are mathematically identical to my conditions (1) and (2) above! Note what this implies. Since the Nash-Harsanyi theory is (intentionally) invariant under separate linear transformations of the utilities, we will always get one and the same physical solution (pie distribution) for *all* admissible choices of the utility scales, including the 'true choices' which correspond to the interpersonally calibrated representations of the utilities. Hence the Nash-Harsanyi theory and the proposed manna theory are solution equivalent, as asserted above. The same type of reasoning can be used to demonstrate the equivalence of our manna theory and the Kaneko-Nakamura theory of social choice.[9]

This result is significant for several reasons. First, it establishes an interesting link between *game theory* on the one hand and *ethics* on the other hand. Let me explore this, and in doing so attempt to provide a new interpretation of the Nash-Harsanyi theory. Shapely (1969) aptly observed that bargaining by its very nature tests the players' relative intensity of desire for what is at stake. Hitherto it has not been possible to make precisely clear what this observation amounts to. The reason why this is so is that virtually all discussions (including Shapley's) take place within the confines of an invariant theory which rules out interpersonal comparisons. Our analysis above puts things in a different light.

We can suppose that we *do* know the true, interpersonally calibrated scales of i and j. With these in hand we can reason that in a given pure bargaining game the players will reach an equilibrium which will distribute utility (fulfill needs) in strict proportion to relative needs. Specifically, we might expect a utility payoff satisfying our equilibrium conditions (1) and (2) above. Finally, we note the 'solution equivalence' between this theory and the Nash-Harsanyi theory. We have in effect used the assumption of interpersonal comparisons to sketch the basis for an alternative account of the Nash-Harsanyi theory.[10]

The fact that the Nash-Harsanyi theory is invariant under separate affine transformations of the utilities can be interpreted in the present context as follows. We can construct two games A and B with the property that the players in A have interpersonally calibrated utility scales which are non-trivial affine transformations of those of the players in B. Hence the distribution of 'needs' in A is *different* from that in B. However, due to Nash invariance, the physical payoff (distribution of pie) will be the *same* in A and B. Is this troublesome? Not at all. For the bargaining theory I have sketched takes the distribution of utility – not of pie – as fundamental. And clearly the distribution of utility awarded by the Nash-Harsanyi theory in games A and B *will be different* – with the difference mirroring the difference in the distribution of needs within the two games. (See also Brock, 1979.)

A second interesting consequence of the Correspondence Theorem lies in its implications for abstract social choice theory. In a recent article reviewing some interesting developments in axiomatic social choice theory, Arrow (1978) argues that the use of interpersonal comparisons is problematic in constructing a satisfactory theory of social choice. On the other hand, he feels that a satisfactory welfare function will possess certain continuity properties which presuppose cardinal utility. Apparently, therefore, he would like a theory which makes use of cardinal utility but which dispenses with interpersonal comparisons, at least at an operational level. The theory I have proposed would seem to meet these desiderata. So does the Kaneko-Nakamura theory.

III.B. *A Comparison With Other Ethical Theories*

I shall now offer a few comments about the role of the concept of relative need in the utilitarian and Rawlsian theories. Thereafter in III.C. I shall contrast the concept of impartiality embodied by my theory with that of the Rawls-Harsanyi Veil of Ignorance concept. In so doing, I shall offer a fresh criticism of Bayesian rationality postulates in ethics.

Consider from a utilitarian standpoint the dilemma of a parent who has to give a present to either of two children. In the case depicted in Figure 5a below, the present – and *all* the utility at stake – will go to child i whose need is ever so slightly greater than j's need. In this type of situation it can in some sense be asserted that utilitarianism sponsors allocation in accord with – but not 'in proportion to' – relative need.

Next consider a pie division problem giving rise to the prospect space of Figure 5b. Utilitarianism here will award slightly more utility to child i than

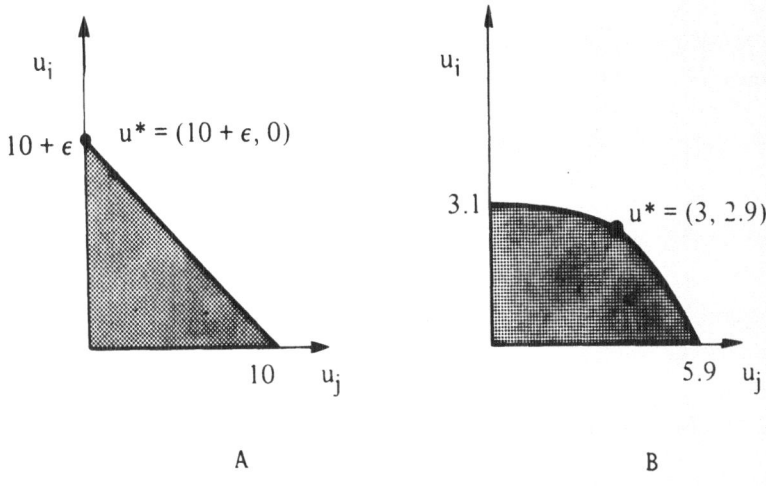

Fig. 5a–b.

to child j. But can we now claim that i is needier than j? Not really, for if all the pie went to i he would only get about one half of the utility that j would get. Finally, in cases like in Figure 5b where corner solutions do not occur, the utilitarian solution clearly will occur at the point of tangency between the iso-utility contour of the utilitarian welfare function – namely the straight line with a $-45°$ slope – and the Pareto boundary of the set U. At this point, the needs of the players are equally urgent. This of course is merely a *local* property of the solution in this case (compare Figure 5a), and it contrasts with the *global* situation in Figure 5b which is not one of equal needs. As a result of these observations, I do not believe that utilitarianism can meaningfully be held to distribute utility in accord with any coherent theory of relative need.

The concept of relative needs plays an important role in Rawls' derivation of the maximin principle. However, Rawls is not always clear, and for my present purposes I shall draw upon Strasnick's (1975) social choice theoretic axiomatization of the maximin principle to criticize the role of needs in maximin-type theories. Strasnick's fundamental theorem is that if ordinal interpersonal comparisons are admitted, and if the social choice function satisfies certain plausible axioms of independence, impartiality, and unanimity, then it must be either a maximax or a maximin decision procedure. Which of these two welfare functions will obtain depends upon which of two parts of a certain lemma we *choose* to adopt. The lemma is called the *extended decisiveness lemma* and states: For all x_i, y_i, y_j, x_j either

$$\text{(a)} \quad x_i > y_j \rightarrow xP_iy > yP_jx; \quad or$$

$$\text{(b)} \quad y_i < x_j \rightarrow xP_iy > yP_jx.$$

Here x and y refer to alternative social states. i and j index the citizens. x_i refers to the position of being i in state x, etc. Finally P is the standard social preference relation. In commenting upon the choice that we must make between part (a) and part (b) of his lemma, Strasnick says:

> ... we must choose whether one individual's preference will have a greater priority than another's whenever he would be left worse-off than the other if his preference were frustrated — or whenever he would be left better-off if his preference were satisfied. Ultimately, this choice must depend on which component of the individual's preference must determine the status of his preference's priority, — his preferred state, or his non-preferred state. Since the individual's condition in his non-preferred state is associated with the nature of his needs, the maximin SCF may be conceived as preferring the same state as the individual with the greatest needs. If we believe that the status of an individual's needs are relevant to the social choice, then we should be prepared to adopt as one of the constraints on the SCF the condition ruling out the legitimacy of part (a) of the extended decisiveness lemma. This would, in fact, be a natural condition if we held the task of social choice to be the distribution of scarce resources among conflicting needs — a task for which the maximin SCF is, some would think, well suited.

I have two problems with Strasnick's insightful account of this matter. First, it seems peculiar to define need in terms of a person's non-preferred outcome. Game theoretic reasoning suggests why this is so. Both Strasnick and I take people's preferences — i.e. their utilities — as fundamental. To speak of a person's being in his worst off state can be translated into his being awarded the utility payoff corresponding to that social state. Now in a constitutional choice problem, it surely seems reasonable — and it is physically possible — for the participants to adopt a jointly randomized strategy if need be. The ability to achieve an *efficient outcome* (in utility space) by this means insures from the outset that *no one will receive his lowest possible utility level*. This being the case, it is not clear why we would ever wish to conceptualize needs in terms of how badly off the worst off person could be. Should we not rather identify needs with the improvements in everyone's well-being which adoption of a constitution will sponsor? I.e. with what is in fact at stake in the decision problem? This is of course my own approach to the matter.

My second problem with Strasnick's approach is the requirement within his set-up for someone to have to have 'preference priority', that is, to be an Arrowian dictator. Analytically this requirement is a natural consequence of a set-up which rules out cardinal utility, as Arrow (1978) and Sen (1977) have pointed out in different ways. If cardinal utility is needed in order for the SCF to possess the continuity properties which permit social welfare to be defined as a reasonable 'balance of advantages' among *all* citizens, then enter cardinal utility. Both the utilitarian theory and my own theory are cardinal in nature.

III.C. *On the Concepts of Realtive Needs and Impartiality*

I shall now conclude this section with a brief discussion of two alternative concepts of impartiality, and their relationship to the concept of allocation in accord with relative needs. Let us suggest that there are two somewhat related but essentially different concepts of a decision which is impartial in an ethical sense of the term. First there is the concept of an ethical decision as a *rational* decision that is made by a rational person who *is* impersonally situated. This is of course the concept of impartiality adopted by Harsanyi (1953) and Rawls (1971).

Second, there is the concept of an ethical decision as the decision which will be made by a person who is *impartial* in the specific sense that he will *only* take *ethically significant information* into account in reaching his decision. Specifically, in a manna distribution problem, he will take account of the players' differences in needs – but that is all. He will not be *biased* by considerations of how attractive or intelligent they are, much as these other considerations will affect his behavior when he is *not* acting in an impartial manner. This second concept of impartiality is consistent with my manna theory. It is also fully consistent with the tradition of *impartial sympathetic humanism* which Harsanyi himself (1958) finds so appealing.

Which of these two concepts is the more appropriate on which to found an ethical theory? In answering this question, I wish to join Harsanyi (1961) in insisting on the importance of separating questions of *ethics* from questions of *rational behavior*. The two are very different, and must not be confused. And yet, they have been confused in the Rawls-Harsanyi concept of an ethical decision as a decision that an impersonally situated rational individual would make. Let me expand upon this point.

The purpose of an ethical theory of distributive justice is to characterize equity, not to serve as the vehicle for an application of rational choice theory. However, if rational choice theory can increase our understanding of equity, then so much the better. But the link between ethics and rational choice theory must be made very clear, and the proper role of rational choice theory in serving ethics must be *demonstrated*, not assumed. Specifically, the decision to use certain axioms of rational choice theory as a foundation for ethics must be justified. And such counterexamples as arise from doing so must be explained. Harsanyi (1975a) has criticized Rawls on both these grounds, and I propose to sketch a brief criticism of his own theory along similar grounds in defense of my own theory.

First, I shall discuss the legitimacy of Harsanyi's use of certain rational choice axioms in an ethical theory. In a recent (1977b) paper, Harsanyi gives a characteristic defense of the three axioms he originally introduced in his landmark 1955 paper:

Axiom (a) (Individual rationality) is an obvious rationality requirement. So is also axiom (b) (Rationality of moral preferences): it expresses the principle that an individual making a moral value judgment must be guided by some notion of social interests (public interest), and indeed, must act as a guardian of these social interests; yet when anybody acts as a guardian of social interests, he must follow, if possible, *even higher standards of rationality* than a person who is merely pursuing his personal interests. Thus if rationality requires that each individual should follow the Bayesian rationality postulates in his personal life as postulate (a) implies, then he must even more persistently follow these rationality postulates when he is making moral value judgments. (Harsanyi, 1977b, p. 9; emphasis added.)

We are told that a person who wishes to make ethical decisions, that is, decisions which are in the social interest, must follow the highest possible standards of rationality[11]. But neither here or elsewhere are we told why this is so. Why should a person such as a public official who is charged with acting in an ethical and fair manner follow standards of *rationality* as opposed to compelling standards of *equity* such as 'To Each According to His Needs?'. This is never explained.

Now if rational choice arguments such as Harsanyi's contributed to the characterization of an intuitively compelling theory of equity then, as I said above, so much the better. But as I have already argued in Section III.A. above, the utilitarian theory is not congruent with our moral intuitions concerning the concept of distribution according to relative needs. Indeed, utilitarianism gives rise to counterexamples such as that appearing in Figure 5a above where a person who has an ever so slightly greater need for pie than another person receives *all* the pie (and hence all the utility) at stake.

In the foregoing discussion I have focused on Harsanyi's deductive, axiomatic derivation of utilitarianism. I chose this model because it provided me the opportunity to express certain doubts about the role of rationality postulates in ethic proper. But Harsanyi has another model, a constructive model in which he derives utilitarianism as the solution to a single person decision problem under uncertainty (e.g., Harsanyi, 1953). In this paper, Harsanyi uses rational choice theory once again. But his use of it here differs from that above. He simply *defines* an ethical decision as an impartial one, where by impartiality he means *impersonal rationality*. The thrust of my above remarks carries over to the present context. Once again, to the extent that rational choice theory is used to derive an ethical theory that is neither needs-respecting nor free from serious counterexamples, there is a problem. For as I have already said, rational choice theory should only enter into ethics to serve ethics.[12] Above and beyond this, there does exist an alternative concept of impartiality which conforms quite well to our instincts about allocation in accord with relative needs. This alternative concept is embedded in my own theory.

IV. ALLOCATION IN ACCORD WITH RELATIVE CONTRIBUTION

In Section II I suggested that full distributive justice would entail the use of the contribution norm in the Stage II decision problem, namely the regime G_c in which people actually choose and live out their life plans. I shall now discuss very briefly how game theory can provide an analytical basis for the ambiguous concept of allocation in according to relative contribution.[13]

Most scholars would probably agree that economic theory has helped illuminate the concept of relative contribution. According to the marginal product theory of general equilibrium analysis, each factor (specifically, labor) is paid a dollar amount equal to its dollar contribution to the organization employing it. There are two fundamental problems with this doctrine which prevent its application in an ethical context. First, the theory only holds in the context of perfectly competitive economies. Our regime G_c may well have an economic system of some sort associated with it; but it will surely entail social systems other than the market, e.g., the family and the polity. Second, the economic theorem states that the dollar value received will equal the dollar value of contribution. But we are interested in *utility*, not dollars, and since people do not all have the same marginal utility for dollars we cannot identify money and utility. Game theory now comes to the rescue and alleviates *both* these problems.

Our basic tool will be the concept of the *non-transferable utility Shapley Value* of a game, henceforth referred to as the value. This solution concept was introduced by Harsanyi (1963) and refined by Shapley (1969). The Shapley Value payoff to player i in an arbitrary n-person cooperative game is

$$(3) \qquad \{u_i = \sum_S \nabla[v(S) - v(S - \{i\})]\}_\lambda \qquad S \subset N$$

where λ is the n-vector of equilibrium 'game weights' (defined below); ∇ is the generalized expectation operator $[(s-1)!(n-s)!/n!]$ in which s is the cardinality of a given coalition S of players, n is the cardinality of the player set in the game; S denotes an arbitrary coalition; N is the player set; and $v(S)$ is the so-called *characteristic function* of the game. Loosely this is a real-valued set function which specifies the 'worth' of any coalition $S \subset N$. More formally, $v(S)$ is the *sum* of the utility payoffs to the s members of S when the coalition plays its 'optimal' strategy in opposing the complementary coalition N/S.

To motivate the concept of the Value of a game, let us first consider the expression which lies within the outer brackets of (3). That is, ignore the entity λ. We can see here the sense in which the Shapley Value of a game provides a measure of a player's contribution to the players in the game. The expression within the inner brackets is clearly a measure – in utility units – of the contribution that player i makes to the coalition S. Now clearly this utility contribution will depend upon S and upon the order in which the various possible coalitions S form. This is the reason for the operator ∇. This expectation operator stipulates that in calculating the contribution of i to the various possible coalitions, all orders of coalition formation are assumed equally likely. The Shapley Value is then defined as the sum of i's contribution to all possible coalitions S, averaged over all possible orders of coalition formation. Clearly the player who receives as his payoff from a game his Shapley Value of the game can be said to be receiving his net contribution (of utility). In short, the Value respects the contribution principle.

In all but a very few special cases (e.g., the case of games with 'side-payment') the value of a game will not exist. However, the Value will *always* exist provided that each person's utility scale is *weighted* by an equilibrium utility weight. Thus the symbol λ appears outside of the outer brackets. Determination of the 'proper' vector of weights is a complex mathematical question I shall not discuss. However, I shall briefly discuss the *meaning* of the Value in light of the need to multiply each person's utility scale by an appropriate weight.

In order to understand the meaning of the Value in the presence of the utility weights, it will be helpful to introduce the assumption that the utility functions of the players have been interpersonally calibrated with respect to both unit and scale. This assumption is not at all necessary for *game theoretical* analysis. However, I believe it is necessary if we are to come to an understanding of the Value suitable for *ethical* analysis. Suppose now that the game in question has been solved for its Value. Suppose moreover that for the particular game in question, $\lambda_1 = \lambda_2 = \cdots = \lambda_n$. In this very unusual case, we know from (3) above that the unweighted utility payoff to player i will be equal to his average net contribution to N as measured in unweighted and interpersonally calibrated utility units. The reason for this is that we can set $\lambda_i = \lambda_j = 1$ with no loss of generality. In this case, the meaning of the Value of the game is crystal clear. In general, however, we are confronted with a situation where the Shapley Value formula appearing within the outer brackets of (3) will only obtain if each person's 'true' (i.e., interpersonally calibrated) utility function is *rescaled* by his utility weight. It can be shown (Brock, 1979) that the equilibrium utility weights are reciprocals of the coefficients that represent the differences among the players in relative need for what is at stake above and beyond the threat payoff of the game.[14] In short, the Shapley Value awards player i a reciprocal-needs-weighted utility payoff equal to his average contribution of reciprocal-needs-weighted utility to the other players.

The implications of all this are interesting for ethical theory. Two avenues are open to us. First, we can be quite loose in our interpretation of the contribution norm and hold that if the solution to a given competitive decision problem is the Value of the problem, then each person is 'receiving' what he has 'contributed', and the contribution principle is being respected. Second, we can take a much more puristic stance and assert that the *only* situation where the contribution norm is being respected is the situation where the utility weights are all equal (relative to the set of interpersonally calibrated utility scales). For my present purposes, I shall adopt the first interpretation here, though not without some hesitation.

For the purposes of our ethical theory, there is one more point to be made about the Shapley Value concept. This concerns its realization in a surprising number of economic and political models of competitive behavior. Harsanyi (1963) formulated a pluralistic bargaining model of social behavior, and showed that his bargaining solution would always be a Shapley Value of the underlying game. Aumann (1975) showed that in certain classes of exchange economies, competitive market trading would realize the Shapley Value. Finally, Aumann and Kurz (1977) have constructed a 'mixed' model incorporating both market trading and legislative (voting) behavior, and have characterized the solution to their model as a Shapley Value. In short, a variety of models of competitive behavior realize the Value, and hence realize distribution in accord with relative contribution. The situation is analogous to that in Section III where we observed that rational behavior in a 'pure' bargaining game will realize distributional equity in the sense of relative needs.

V. REALIZATION OF FULL DISTRIBUTIVE JUSTICE VIA A CONTRACTARIAN MODEL G^*

Our results put us in a position to characterize full distributive justice (as defined in Section II.C above) in terms of the play of a two-stage game G^*. It will probably help the reader here to consult Figure 1. The first-stage game is an n-person pure bargaining game G^c. The disagreement payoff in this game is the payoff d^* awarded in the game G_M. G_M it will be recalled is the default game in which players will participate if they cannot reach unanimous agreement in G^c on the choice of an optimal constitution c^*. The prizes at stake in G^c are all (joint randomizations of) the alternative possible constitutions $c \in C$, or equivalently the regimes $\{G_c\}$ induced by adoption of the various constitutions. The stipulation that G^c be a pure bargaining game ensures that the 'needs principle' is being respected in its proper domain.

The second-stage component game of G^* is simply the regime G_{c^*} chosen as the solution to the stage-one problem G^c. The strategies of G_{c^*} lead to an outcome which is a Shapley Value of the game. For when this happens, the 'contribution principle' is being respected in its proper domain, that is, in the second stage of G^*. As we argued above, it is only in this second stage that the players can meaningfully be said to be making differential contributions to the social product.

By assumption, both the first and second-stage component games of G^* are cooperative games. Indeed, both the Nash solution and the Shapley Value are cooperative game theories. Moreover, each of these solutions awards the players a utility payoff which is an imputation. An imputation is a payoff vector which is both Pareto optimal and individually rational. (A payoff vector is said to be individually rational if it leaves every player better off than he is to begin with.) Finally, since the payoff from the two-stage game G^* is defined as the payoff from the optimal second-stage game G_{c^*}, we know that the payoff of G^* itself is an imputation. Now given the intuitive meaning of a 'contract' as an agreement everyone enters into to improve his situation, – i.e., an agreement awarding an imputation – it seems reasonable to assert that G^* is a *Contractarian* rational choice model. Accordingly, we can summarize our results above with a statement of our

FUNDAMENTAL REALIZATION THEOREM. The ethical concept of full distributive justice can be conceptualized in Contractarian rational choice theoretic terms, and can be realized through a play of a specific two-stage cooperative game G^*.

A formal characterization of G^* and proof of this assertion is found in my companion paper (Brock, 1978). In that paper I have also investigated the problem of political representation. It is shown that there exists an aggregated version of G^* – a game G^{**} played by set of r strategic representatives of the n citizens – which is strategically equivalent to the original game G^* under certain conditions. G^{**} can be viewed as an aggregated Contractarian version of G^*.

Of course, it is not necessary to conceptualize distributive justice in terms of a Contractarian rational choice model. For our earlier discussion makes clear that full distributive justice could as well be realized through an ethical arbitration scheme. Nonetheless, it is significant that a Contractarian model of our theory is available, given the role of Contractarian theories in the history of Western political and moral theory. In this vein, it is interesting to note that the component game G^c (the constitutional choice game proper) of G^* might be viewed as a formal representation of the bargaining game which Rawls originally envisioned when he first introduced his theory of justice in 1957 (see Wolff, 1977, Chapters I–V). Our stipulation that G^c be a *pure* bargaining game might in this context correspond to Rawl's requirement that the players be reasonably equal in power and ability.

Stanford Research Institute and
University of Texas at Austin

NOTES

[1] The author wishes to acknowledge the advice and criticism he has recieved during the development of this theory from Kenneth J. Arrow, John C. Harsanyi, Thomas M. Scanlon, and Lloyd S. Shapley. He is of course responsible for all deficiencies of a technical or conceptual nature. This research has been supported by grants from the Andrew Mellon Foundation, The Aspen Institute for Humanistic Studies, The U.S. Department of Defense (D.A.R.P.A. Contract No. 78-072-0723), and the U.S. Department of Energy (DOE Contract No. EX-76-C-01-2089).

[2] Professor Charles Taylor (1976) expresses the view that these two norms are indeed the fundamental distributive norms. And he cites the inability of existing theories to reconcile these norms as a singular deficiency in moral theory. Additionally, both Nozick (1974) and Wolff (1977) have criticized the Rawlsian and the utilitarian theories for neglecting the question of contribution.

[3] The distinction between manna and non-manna environments was apparently introduced by Robert Nozick (1974, Chapter VII).

[4] To introduce the concept of a disagreement payoff or zero point should not be interpreted to prejudge whether or not a suitable welfare function should depend on the zero point. I discuss this matter further on.

[5] The only concept that contends with 'relative needs' for hegemony here is the concept of impartiality (e.g., Harsanyi, 1953). I shall discuss the relationship between these two concepts in Section III.B.

[6] In the context of a pie division problem where there is a continuously divisible commodity, this case will correpsond to an assumption that both players have *constant marginal utility* for pie.

[7] The concept of impartiality as defined here is very similar to the game theoretic axiom of the Expected Independence of Irrelevant Variables used by J. C. Harsanyi (1977a, pp. 154–157) in his axiomatization of the Nash-Zeuthen theory.

[8] In a future paper, I shall argue that there exists a proper' zero point that is not morally arbitrary in the context of a theory of justice.

[9] Kaneko and Nakamura do not assume interpersonal utility comparisons. They do assume cardinal utility and the existence of a zero point. They postulate
  (i) Invariance of the welfare function under a relabeling of the players and of the social states (Symmetry and Neutrality);
  (ii) Independence of irrelevant alternatives; and
  (iii) Pareto optimality.
They establish that the *only* welfare function satisfying (i)–(iii) is the Nash-Harsanyi function which calls for a maximization of the arithmetic product of the players' utility gains. I shall not discuss this interesting result further in the present paper.

[10] Harsanyi (1977a, 194) has suggested that the invariance of the Nash theory can be used to establish what he calls an *ad hoc* interpersonal comparison. Specifically, we are *free* to rescale the utilities separately such that the game weights a_i, a_j become *equal to unity* in which case the solution (recall Equation (1) above) assumes the form $(u_i - d_i) = (u_j - d_j)$. This is indeed an *ad hoc* description since the utilities will not be meaningfully interpersonally calibrated before or after the rescaling. Our own argument is fundamentally different from Harsanyi's. It illuminates how we can interpret the Nash solution if we were to start off with and retain meaningfully interpersonally calibrated scales – something which it is not necessary to do in the context of the Nash-Harsanyi theory.

[11] Incidentally, even if we accept Harsanyi's assertion of the need for highly rational behavior (as opposed to highly ethical behavior) on the part of an ethically motivated person acting in the social interest, it is not at all clear why the particular rationality postulates that he should follow are the Bayesian postulates that apply to the special case of *individual* decision-making under risk and uncertainty.

[12] Note that when I speak of rational choice theory, I am in effect referring to single-person decision theory, *not* to game theory which enters into my own theory as has been shown above.

[13] Whereas the present paper does provide a thorough account of the concept of allocation according to relative needs, lack of space prevents my giving an equally detailed account of the contribution principle. A detailed treatment of the latter can be found in a companion paper (Brock, 1978).

[14] Aumann (1975) has proven the equivalence of the competitive equilibrium of economic theory with the non-transferable utility Value discussed above. He asserts that because of his result, the market allocates utility in accord with relative contribution. However, he does not provide an interpretation of the weights. Hence his conclusions are ambiguous.

BIBLIOGRAPHY

Arrow, Kenneth J.: 1978, 'Extended sympathy and the possibility of social choice', Philosophia VII, No. 2.

Aumann, Robert J.: 1975, 'Values of markets with a continuum of traders', Econometria 43.

Aumann, Robert J. and Mordecai Kurz: 1977, 'Power and taxes in a multi-commodity market', Israel Journal of Mathematics 27.

Brock, Horace W.: 1977, 'Unbiased representative decision-making: A study of invariance relationships in *n*-person valutaion theory', Proceedings J.A.C.C.

Brock, Horace W.: 1978, 'A new theory of justice based on the mathematical theory of games', appearing in: Game Theory and Political Science, edited by P. Ordeshook (New York University Press, New York).

Brock, Horace W.: 1978a, 'The role of symmetry groups in the representation and interpretation of alternative theories of fair distribution', Proceedings of the Naval Postgraduate School Conference in Mathematics Linkages.

Brock, Horace W.: 1978b, 'The Shapley Value as a tool for the conceptual unification of economics, politics, and ethics', Proceedings of the American Political Science Meetings.

Brock, Horace W.: 1978c, 'A critical discussion of the work of John C. Harsanyi', Theory and Decision 9.

Brock, Horace W.: 1979, 'The problem of "utility weights" in group preference aggregation', Operations Research (forthcoming).

Harsanyi, John C.: 1953, 'Cardinal utility in welfare economics and in the theory of risk-taking', Journal of Political Economy 61.

Harsanyi, John C.: 1955, 'Cardinal welfare, individualistic ethics, and interpersonal comparisons of utility', Journal of Political Economy 63.

Haranyi, John C.: 1958, 'Ethics in terms of hypothetical imperatives', Mind 67, pp. 305–316.

Harsanyi, John C.: 1961, 'On the rationality postulates underlying the theory of cooperative games', Journal of Conflict Resolution 5.

Harsanyi, John C.: 1963, 'A simplified bargaining model for the n-person cooperative game', International Economic Review 4.

Harsanyi, John C.: 1975, 'The tracing procedure: A Bayesian approach to defining a solution for n-person non-cooperative games', International Journal of Game Theory 4.

Harsanyi, John C.: 1975a, 'Can the maximin principle serve as the basis for morality: A critique of John Rawls' theory', American Political Science Review 69.

Harsanyi, John C.: 1977a, Rational Behavior and Bargaining Equilibrium in Games and Social Situations (Cambridge University Press, Cambridge).

Harsanyi, John C.: 1977b, 'Bayesian decision theory and utilitarian ethics', Working Paper CP-404, Center for Research in Management Science, University of California, Berkeley.

Harsanyi, John C.: 1978, 'A solution theory for non-cooperative games and its Implications for cooperative games', appearing in: Game Theory and Political Science, ed. by P. Ordeshook (New York University Press, New York).

Kaneko, Mamoru and K. Nakamura: 1977, 'The Nash social welfare function', forthcoming in: Econometrica.

Nozick, Robert: 1974, Anarchy, State and Utopia (Basic Books, New York).

Rawls, John: 1971, A Theory of Justice (The Bellknap Press of Harvard University, Cambridge).

Sen, Amartya: 1977, 'On weights and measures: Informational constraints in social welfare analysis', Econometrica 45, No. 7.

Strasnick, Steven: 1975, Preference Priority and the Maximization of Social Welfare, Unpublished Ph.D. Thesis, Harvard University.

Shapley, Lloyd S.: 1969, 'Utility comparisons and the theory of games', appearing in: La Decision: Aggregation et Dynamique des Ordres de Preference, ed. by G. Th. Guilbaud (Centre Nationale de la Recherche Scientifique, Paris).

Taylor, Charles: 1976, 'Normative criteria of distributive justice', Unpublished conference paper communicated to the author through the courtesy of Professor Daniel Bell of Harvard University.

Wolff, Robert P.: 1977, Understanding Rawls (Princeton University Press, Princeton).

ALLAN GIBBARD

DISPARATE GOODS AND RAWLS' DIFFERENCE PRINCIPLE: A SOCIAL CHOICE THEORETIC TREATMENT

ABSTRACT. Rawls' Difference Principle asserts that a basic economic structure is just if it makes the worst off people as well off as is feasible. How well off someone is is to be measured by an 'index' of 'primary social goods'. It is this index that gives content to the principle, and Rawls gives no adequate directions for constructing it. In this essay a version of the difference principle is proposed that fits much of what Rawls says, but that makes use of no index. Instead of invoking an index of primary social goods, the principle formulated here invokes a partial ordering of prospects for opportunities.

1. PRIMARY SOCIAL GOODS AND THE INDEXING PROBLEM

In *A Theory of Justice*, Rawls claims as one of the virtues of his theory that it does not require interpersonal comparisons of utility. Instead, the interpersonal comparisons needed for the theory are based on an 'index of primary social goods'. Primary goods are goods useful toward widely disparate ends, "things which it is supposed a rational man wants whatever else he wants" (p. 92).[1] The primary social goods include rights and liberties, powers and opportunities, income and wealth (pp. 62, 92). In Rawls' theory, then, the basis of interpersonal comparisons is overt: the comparisons are to be made on the basis of who gets how much of what.

Why might this be an advantage? To anyone who denies that interpersonal comparisons of utility are meaningful, the advantage will seem obvious: by not invoking such comparisons, the theory avoids a pseudo-concept. Even if such comparisons are meaningful, though, they should perhaps still be avoided in the formulation of a public conception of justice – a conception which is to be used in resolving conflicts of interest over the basic structure of society. For even if such comparisons can be made in principle, they will often be delicate, and the relevant psychological evidence will probably not be compelling. When interests conflict, and delicately based judgments are to be used to adjudicate them, each person's judgments are

Theory and Decision 11 (1979) 267–288.   0040–5833/79/0113–0267$02.20.

likely to reflect his own interests. When that happens, there will be no agreement on how the standards of adjudication apply to the conflict in question. A public conception of justice should set up standards that are easy to apply, and interpersonal comparisons of utility, even if meaningful, do not pass this test (cf. pp. 90–93).

How clear, then, is Rawls' own standard? He speaks of an 'index' of primary social goods, which is to provide a clear standard for interpersonal comparison; my question concerns how that index is to be specified.[2] For it is this index that gives content to Rawls' *difference principle*: that the index of primary goods for the worst-off representative man is to be as high as is feasible (pp. 83, 90–95).

It might be thought that the specific index used does not greatly matter for the content of the difference principle. The index is used to identify the worst-off representative man, and those who are worst off in one primary good are likely to be worst off in all. In that case, all indices will agree on who is worst off, however differently they weigh the various primary goods.[3]

The index is used, though, not only to identify the worst-off representative man, but to compare various alternative arrangements of society from his standpoint: that social arrangement is just which accords the highest feasible index to the worst-off representative man. Now alternative social arrangements may differ vastly in the kinds of rewards they offer: capitalism with an income floor, for instance, might offer the worst-off representative man considerable income with few powers, whereas some alternative might offer him a lower income with more powers. How the index weighs income against powers will determine which social arrangement accords the worst-off representative man the higher index of primary social goods.

Some of the things Rawls says suggest that the index is not to be part of the difference principle itself, which is that "social and economic inequalities are to be arranged so that they are . . . to the greatest benefit of the least advantaged" (p. 83). True, to explicate the phrase 'to the greatest benefit of the least advantaged', we need an index, since the phrase means 'such that the least expected index of primary social goods in the society is as great as is feasible'. Rawls suggests, though, that whereas the difference principle is to be adopted in the original position, which is the first stage of a four-stage sequence of deliberation, the index that explicates it is to be left to a third, 'legislative' stage (1974, p. 642). In the legislative stage, as in the original position, no one knows his own identity, abilities, and plan of life; but in the legislative stage, unlike the original position, "the full range of general economic and social facts" about the particular society in question can be brought to bear (p 199). The difference principle, then, is adopted by parties ignorant of the particular circumstances of their own society, but the index that interprets it is to be constructed after the parties have learned what their particular society is like. At that point, the index is to be constructed "by taking up the standpoint of the representative individual" from the worst-off group,

and asking which combination of primary social goods it would be rational for him to prefer. In doing this, we admittedly rely on our intuitive capacities. This cannot be avoided entirely, however. The aim is to replace moral judgments by those of rational prudence and to make the appeal to intuition more limited in scope, more sharply focused (p. 94).

More, though, needs to be said. The difference principle is of indeterminate meaning until we know how, given the circumstances of any particular society, to construct the index through which the principle applies to that society. The construction cannot come directly from judgments of rational prudence, since what matters is the rational preferences of 'the representative individual' of the worst-off group, and 'the representative individual' is not a person. Rather, statements about 'the representative individual' abbreviate more complex statements about a class of individuals. What is rationally prudent for the representative worst-off individual must in some way be a matter of what is rationally prudent for genuine individuals, or of what would be, under certain circumstances.

Perhaps we are to construct the index for a society by asking what would be rationally prudent for a person who knew that he would start out in the worst-off group in that society, who knew what the society was like, but who did not know particular facts about himself. Before that says much about how to construct the index, though, it must be joined with an account of how it is rational to choose with limited information. For what is rational under limited information has notoriously been a matter of controversy in discussions of Rawls' theory. The difference principle has no clear content until directions are given for constructing the index of primary social goods in terms of which the principle is stated.

2. THE INDEXING PROBLEM FOR INCOME

In all but the last section of this paper, I shall discuss the indexing problem for only one primary good, income. Income closely fits Rawls' description of a primary good as something a rational person wants whatever else he wants. For an income is not an allotment of particular commodities, but rather an opportunity to choose among the most diverse combinations of goods. An income, then, can be used in the pursuit of a wide range of alternative sets of goals.

Restricting the initial discussion to income has a number of advantages. If conceptual problems arise with income, they will presumably remain when other primary social goods are included in the problem. We can perhaps best identify those problems by making simplifying assumptions. On the other hand, solutions to conceptual problems that arise in the case of income may turn out to apply to the general case of disparate primary social goods.

We may, if we wish, think of the special case of income as follows: we are now restricting ourselves to cases in which all primary goods other than income are distributed equally in a fixed amount, and then asking what would constitute maximizing the prospective incomes of those in the worst starting positions. For the sake of even more simplicity, I shall consider at the outset only cases of certainty, and suppose that we want to maximize the income of the worst-off person.[4]

Why might there be an indexing problem for incomes? An income, as I have said, amounts to a choice among diverse combinations of goods. Now although such a wide range of choices might be offered in other ways, the term 'income' suggests a particular way in which such a choice can be offered: one is faced with a quantity of income and a system of prices, and one may choose any combination of consumption goods the total price of which does not exceed one's income. For the sake of simplicity, I shall consider only such pure income-price systems at this point.

The indexing problem for incomes is this. Alternative economic policies may produce different relative prices. It may be that given one economic policy, the worst-off person would face one income-price combination, and given another policy, the worst-off person would face another income-price combination. More than one person may be worst-off, and those people may not agree on which income-price combination is preferable. In that case, how is it to be settled which policy leaves the worst-off people best off?

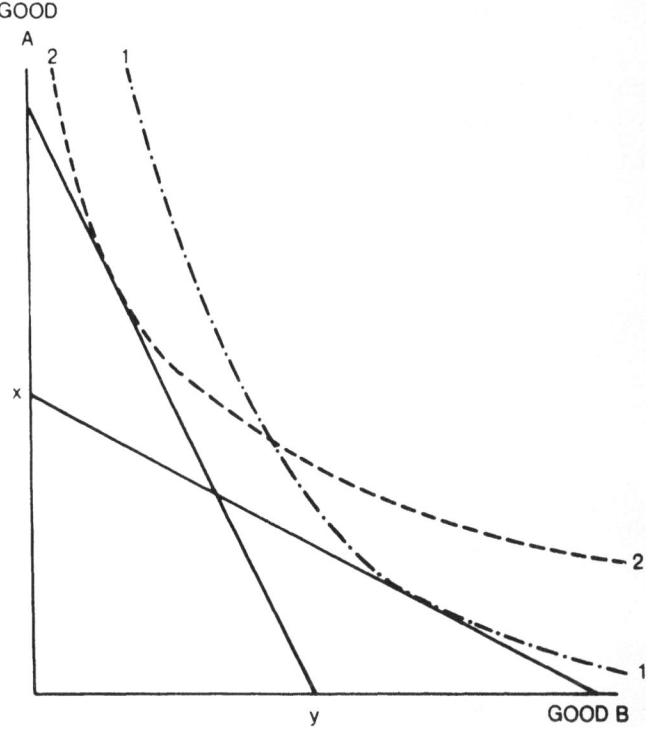

Fig. 1.

The situation can be represented graphically. Suppose there are only two commodities, *A* and *B*. Any point in a two-dimensional graph then represents some combination of goods *A* and *B*, or *commodity bundle*. The purchasing power of an income in the face of certain prices can be represented by a *budget constraint*, such as *x* in Figure 1: given that income and those prices, the person can purchase any commodity bundle on his budget constraint. He can also purchase less of any commodity. The set of commodity bundles he can purchase is his *budget set*.

I shall assume that each commodity is a good, in the sense that other things equal, each person prefers more of it to less. It follows from this assumption that a person's most preferred bundles lie on his budget

constraint. It also follows that given fixed prices, everyone agrees in preferring greater income to less.

Disagreement begins when alternative price ratios are considered. In
Figure 1, the preferences of two people, 1 and 2, are represented by indifference curves in the space of commodity bundles. x is a budget constraint
for someone who faces a high price for good A and a low price for good B;
y is a budget constraint for someone who faces a low price for A and a high
price for B. Person 1 prefers x to y and person 2 prefers y to x. An example
of the indexing problem for incomes, then, is this: if one economic policy
would face the worst-off members of society with constraint x and another
would face them with constraint y, which is the more just by the standards
of the difference principle? Which policy, that is to ask, leaves the worst-off
members of society better off?

Is this really a problem of justice? Questions of justice, on Rawls' view,
concern the basic structure of society, and the details of relative prices
do not seem part of the basic structure. Justice in prices will be procedural:
once the basic structure is just, the prices which emerge from the procedures
it sanctions are just simply because they emerge from those procedures
(cf. pp. 86–9, 304–9).

With all this granted, however, the indexing problem remains one of
justice. Whether the basic structure itself is just will depend in part on the
incomes that could be expected to emerge from it, and different arrangements of the basic structure can be expected to produce different price
ratios. Among the basic economic decisions to be made are the degree to
which prices should be administered rather than set by markets, whether
necessities such as food and shelter should be subsidized, whether rec-
reational and cultural activities should be subsidized, and so forth. All these
decisions involve making some commodities cheaper than would the market,
and any taxes used to finance subsidies will make some commodities more
expensive than would the untaxed market. Decisions about the basic economic structure of society affect price ratios as well as incomes, and we need
some way of judging the alternatives by their expectable upshots.[5]

3. A MINIMAL DIFFERENCE PRINCIPLE

I want to propose not a way of constructing an index of budget sets, but a
weak version of the difference principle that does without an index. The

proposal in this section is weak and preliminary, and later I shall consider how to strengthen it so as to rule out as unjust some economic states which this weak preliminary version admits as just. Even in this preliminary version, though, the difference principle has enough strength to be in conflict with the norms of efficiency: it may be that no Pareto efficient economic state is just by the difference principle, even in this weak formulation.

I begin with some definitions and notation. Since the range of choice an income represents is determined by prices, we must represent an individual's situation by giving not simply his nominal income, but the prices he faces. An *individual state*, then, will consist of an *income* and a *price vector*. The *income* is a non-negative number, and the *price vector* is an assignment of a non-negative number to each *commodity* in a finite non-empty set $\mathscr{C}$. A *social state* consists of a distribution and a price vector, where a *distribution* is an assignment of an income to each *person*, or member of a finite non-empty set I. Feasible social states[6] will be represented by italic, bold-face letters w, x, y, z. Individual states will be represented by italic letters w, x, y, z, with or without subscripts. People will be represented by i, j, and k, and where w is a social state, w_i will be the individual state of person i in w: the individual state, that is, consisting of the income of i in w and the price vector of w.

Since where prices differ, incomes may not be comparable in any obvious way, we might do well to start with comparisons only of individual states with the same price vectors. Such individual states will be called *directly comparable*. Let $x > y$ iff x and y have the same price vector and the income of x is greater than that of y; in that case, we shall call x *directly better* than y. The relation $\succ$, then, will be a *strict partial ordering* of individual states: transitive and irreflexive.

Although $\succ$ gives few comparisons, we can use it to construct a very weak comparison of social states by a version of the difference principle. According to the difference principle, a social state x is more just than social state y iff the worst-off person in x is better off than the worst-off person in y, and a social state x is just iff there is no feasible social state that is more just than x. To say that x is more just than y, in other words, is to say that there is someone such that everyone is better off in x than he is in y. Say, then, that social state x is *directly more just* than social state y, or $x \mathscr{J}^* y$, iff x and y have the same price vector and x has a higher minimum income than y; in other words,

$$x \mathscr{J}^* y \quad \text{iff} \quad (\exists j)(\forall i)x_i > y_j.$$

A permissive version of the difference principle is that a social state x is just iff there is no feasible social state which is directly more just than x; we say, then, that x is *minimally just* iff $\sim (\exists z)z \mathscr{J}^* x$. A social state is minimally just, in other words, iff there is no feasible social state with the same prices and a higher minimum income.

Now even the very restricted standard of comparison expressed by $\mathscr{J}^*$ is incompatible with the weak Pareto principle: that if everyone prefers social state x to social state y, then x is better than y. The standard is incompatible with the Pareto principle not only in the sense that we may have x unanimously preferred to y without having x directly more just than y. It is incompatible in that if we try to combine the comparisons made by $\mathscr{J}^*$ with comparisons made by the Pareto principle, we may have cycles.

Where each person i has an ordering P_i of individual states, we might set the following two conditions on a relation $\mathscr{J}$, to be read 'is better than' or 'is more just than'.

Unanimity: For any x and y, if $(\forall i) x_i P_i y_i$, then $x \mathscr{J} y$.

Minimal Difference Principle: For any x and y, if $x \mathscr{J}^* y$, then $x \mathscr{J} y$. We can find patterns of preference such that Unanimity and the Minimal Difference Principle cause $\mathscr{J}$ to cycle, and we can do so in the case of two goods and two people. A case is given in Figure 2; the idea behind it is this. Let person 1 prefer good A and person 2 prefer good B. Start 1 and 2 out in state z with equal incomes. Produce a Pareto improvement as follows. Raise the price of B and lower that of A in such a way as to please 1 and displease 2. Drop 1's income, but only slightly, so that 1 is still better off in the new state than he was in z. Raise 2's income enough to overcompensate him for the price change. Call the new state y; then y is unanimously preferred to z. Now, leaving prices unchanged, form state x by returning 1 and 2 to their original incomes. That raises 1's income and lower 2's, producing, by the Minimal Difference Principle, a more just state. Thus x is more just than y by the Minimal Difference Principle, and y is unanimously preferred to z. Yet x differs from z only in that prices are changed to 1's advantage and 2's disadvantage. If a combination of Unanimity and the Minimal Difference Principle shows x to be better than z, then it is clear from the symmetry of the case that an analogous reverse argument will show that z is better than x.

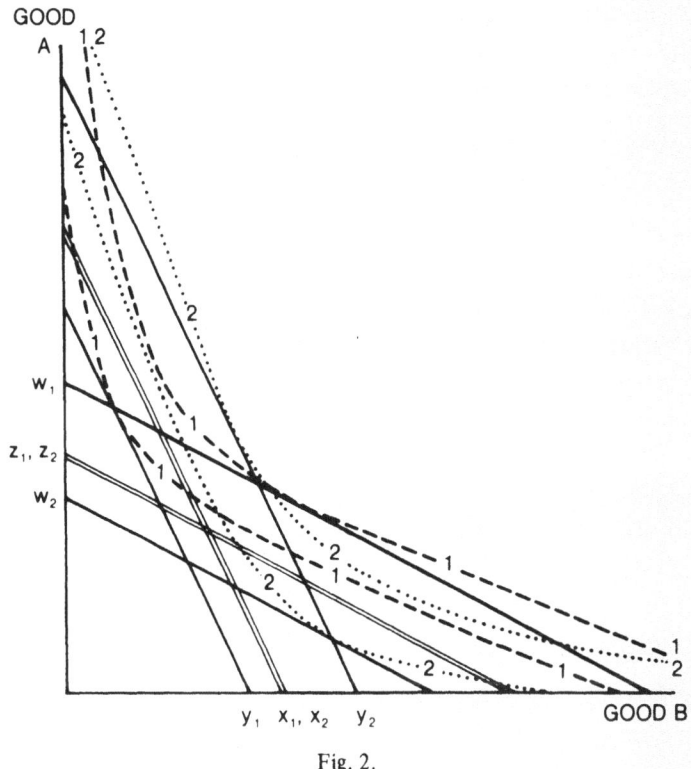

Fig. 2.

In Figure 2, y is unanimously preferred to z, x is more just than y by the Minimal Difference Principle, w is unanimously preferred to x, and z is more just than w by the Minimal Difference Principle. The difference principle, even in its most weak and unproblematic form, cannot be reconciled with the weak Pareto principle.

The point might be put another way. In the case we have been considering, if w, x, y, and z are the only feasible social states, then only w and y are Pareto optimal, but according to the Minimal Difference Principle, at most x and z are just. Now any principle that might reasonably be called a version of the difference principle will at least say what the Minimal Difference Principle says: that prices equal, the economic system with the higher minimal income is more just. For the very idea of the difference principle is to use overt criteria such as income for making interpersonal comparisons, and to judge economic systems by how well off, by those overt criteria, they make the worst-off members of society. What we have shown, then, is that any faithful explication of the difference principle will yield a criterion of economic justice which is incompatible with the Pareto principle.

4. STRENGTHENING THE PRINCIPLE: SOME PITFALLS

I turn now to strengthening the difference principle so as to rule out more economic systems as unjust. At this point, I shall make no attempt to reconcile the difference principle with the Pareto principle, for we have seen the two to be irreconcilable. Later, I shall discuss whether the conflict between the two principles is a bad thing, and how the difference principle could be modified to avoid the conflict if we wanted to do so.

Note first a pitfall. Since people differ in their preferences among individual states, we might want to compare not simply individual states, but people in individual states. Let a pair $\langle i, x \rangle$ consisting of a person i and an individual state x be called a *personal state*. Perhaps instead of ranking individual states, which in effect are simply budget sets, we should rank personal states. That will allow us to take into account the preferences of the people involved.

Now it is central to the difference principle that within a single social state, interpersonal comparisons are to be by income. For any individual state x, let $I(x)$ be the income in that state. Let $\langle i, x \rangle B \langle j, y \rangle$ mean 'person i is better off in individual state x than is person j in individual state y'. Then we must require the following.

Interpersonal Comparison by Income (CI): For any social state x and people i and j, if $I(x_i) > I(x_j)$, then $\langle i, x_i \rangle B \langle j, x_j \rangle$.

An apparent advantage of considering personal states rather than mere budget constraints is that we can now make intrapersonal comparisons by consulting the preferences of the person involved.

Intrapersonal Comparisons by Preference (CP): For any person i and individual states x and y, if $xP_i y$, then $\langle i, x \rangle B \langle i, y \rangle$.

Conditions CI and CP, though, will sometimes force the relation B to be cyclic. In Figure 3, we have

(1) $\langle 1, x_1 \rangle B \langle 2, x_2 \rangle$ by   CI;

       $\langle 1, y_1 \rangle B \langle 1, x_1 \rangle$ by   CP;

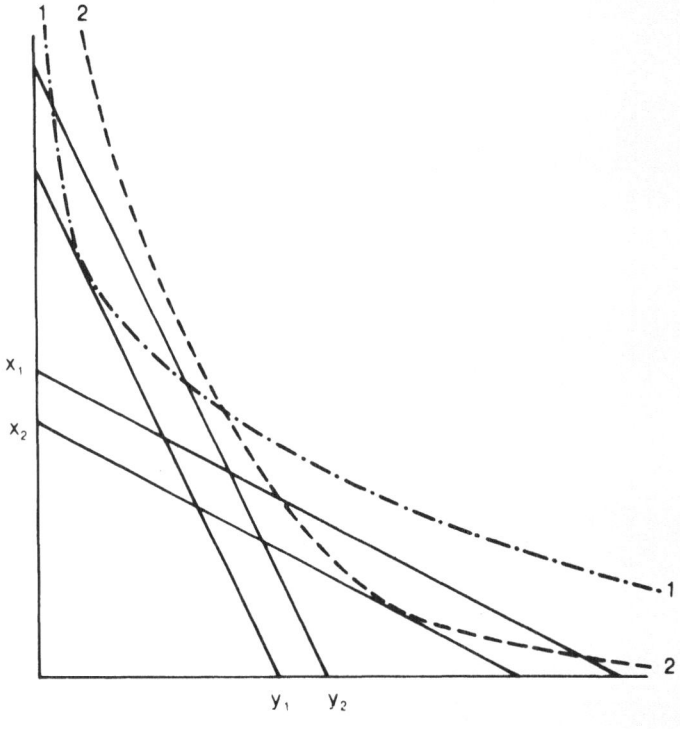

Fig. 3.

(2) $\langle 2, y_2 \rangle B \langle 1, y_1 \rangle$ by   CI;

        $\langle 2, x_2 \rangle B \langle 2, y_2 \rangle$ by   CP.

Thus B is cyclic.

We have not shown that where B is cyclic, a $\mathscr{I}$ defined from B must be cyclic. Nevertheless, the ease with which the relation of being better off can be made to cycle is grounds for caution. How should we proceed? One approach to making limited judgments of equity without pschologically based interpersonal comparisons is through a concept of 'envy'. The concept as originally formulated applied to bundles of goods: person i envies j's bundle of goods if he would rather have it than his own. The same consider-ations could be applied to personal states. We could say that person i in state x_i *envies* person j in state y_j if he prefers y_j to x_i – if, in other words, $y_j P_i x_i$. Now there can easily be cases where $y_2 P_1 x_1$ and $x_1 P_2 y_2$, as in

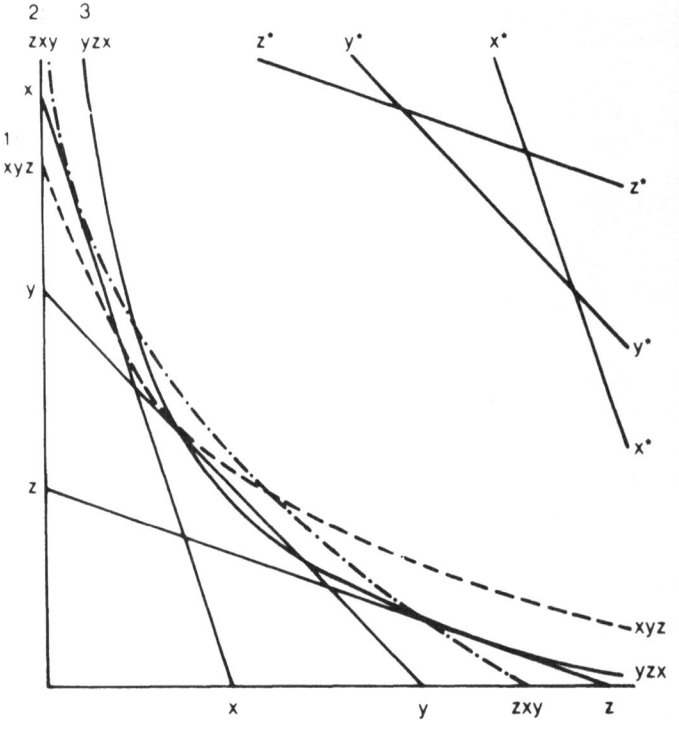

Fig. 4.

Figure 1 with $x_1 = x$ and $y_2 = y$. Perhaps the relation we should use is this: We say that i is *definitely better off* in x_i than is j in y_j if both i and j prefer x_i to y_j. I shall write this $x_i P_{ij} y_j$.

That suggests the following version of the difference principle. We do not hope for a complete weak ordering of personal states, but only for a partial ordering. B will be the relation *is definitely better off than*, and it holds between personal states. We have $\langle i, x \rangle B \langle j, y \rangle$ iff xP_iy and xP_jy. We define the relation *x is more just than y*, or $x \mathscr{J} y$, as follows:

(4) $x \mathscr{J} y$ iff $(\exists j)(\forall i)\langle i, x_i \rangle B \langle j, y_j \rangle$.

Social state x is *just* iff no feasible social state is more just than x.

Now this new $\mathscr{J}$ too is cyclic. To see this, note first that even when preferences are well-behaved in the usual ways, there will be triples of income-price situations such that any preference ordering of them is possible. Moreover, where x and y are individual states, it will be possible to find an individual state x^* that is directly comparable to x such that x^*P_iy. That is, it is always possible to raise someone's income enough to compensate him for an unfavorable shift in prices. Hence we may suppose we have the following orderings of individual states: 1: xyz; 2: zxy; 3: yzx, where everyone prefers each of x^*, y^*, z^* to each of x, y, z, and moreover, x^* is directly comparable to x, y^* to y, and z^* to z. The situation is as shown in Figure 4. Let the social states be

$$x = \langle x, x, x^* \rangle,$$

$$y = \langle y, y^*, y \rangle,$$

$$z = \langle z^*, z, z \rangle.$$

Then $x \mathscr{J} y$, because xP_1y, $xP_{12}y$, and $x^*P_{13}y$. Like arguments show that $y \mathscr{J} z$ and $z \mathscr{J} x$, yielding a cycle.

5. A WAY OUT

The idea of this proof was to exploit an appropriate cyclic majority among three individual states. 1 prefers x to y, and so he prefers everyone's state in x to his in y. 2 also prefers x to y, and so prefers his state in x to 1's in y. Even though 3 prefers y to x, he is made rich enough in x to prefer his state in x to 1's in y. In that sense, x makes everyone better off than someone is made in y.

Since this relation is cyclic, we need to strengthen the difference principle as applied to $\mathscr{J}$ so as to make fewer comparisons. Note that in the above example, although 1 and 2 agree that 2 is better off in x than 1 is in y, 3 does not: 3 prefers y to x. Perhaps, then, we should consider i better off in x than j is in y only if everyone prefers individual state x_i to individual state y_j. Our standard now makes no mention of the individuals involved, and so we can consider it a method for comparing individual states rather than personal states. We can now say xBy iff $(\forall i)xP_iy$; in that case, we say that *one is unequivocally better off* in individual state x than in individual state y. x is *more just* than y, or $x \mathscr{J} y$, iff there is some individual j such that one is unequivocally better off in anyone's state in x than one is in j's state in y.

Formally put, we have $x \mathcal{F} y$ iff $(\exists j)(\forall i)x_i By_j$. The *Difference Principle* will say that x is just iff there is no feasible state z such that $z \mathcal{F} x$.

$\mathcal{F}$ is a strict partial ordering; that is, transitive and irreflexive. The simple lemma behind this claim will be useful in later discussion, and bears explicit statement. Note at the outset that B as defined here is a strict partial ordering, since it is simply a relation of unanimous preference among individual states.

LEMMA: Let B be any strict partial ordering of individual states, and define $x \mathcal{F} y$ as $(\exists j)(\forall i)x_i By_j$. Then $\mathcal{F}$ is a strict partial ordering of social states.

Proof: $\mathcal{F}$ is irreflexive, because $x \mathcal{F} x$ means $(\exists j)(\forall i)x_i Bx_j$, which entails that for some j, $x_j Bx_j$, contradicting the irreflexivity of B. To see that $\mathcal{F}$ is transitive, suppose $x \mathcal{F} y$ and $y \mathcal{F} z$. Then there are a j such that $(\forall i)x_i By_j$ and a k such that $(\forall i)y_i Bz_k$. Therefore, $y_j Bz_k$, and since $(\forall i)x_i By_j$, by transitivity of B, we have $(\forall i)x_i Bz_k$. Therefore $x \mathcal{F} z$.

Since $\mathcal{F}$ is a strict partial ordering, there will always be at least one feasible social state which is just according to the Difference Principle. We have seen that in some cases, no feasible social state which is just according to the Difference Principle will be Pareto optimal, for the Minimal Difference Principle allowed all states as just that the Difference Principle does, as a check of the criteria will show, but it was Pareto-incompatible.

What the Difference Principle in its present form says is this. A social state x is just unless for some feasible social state z, everyone agrees that he would rather face the prices in z with the income of the poorest person in z than face the prices of x with the income of the poorest person in x. We have, then, a consistent version of the Difference Principle as restricted to income. That version is, I suspect, as strong as it can be made without introducing either an element of arbitrariness, or interpersonal comparisons of happiness, strength of preference, or the like.

6. PARETO COMPATIBILITY

How should we regard the Pareto-incompatibility of the Difference Principle? Does the joint inconsistency of the Difference Principle and the Pareto Principle disquality the Difference Principle as a principle of justice? I think not. Remember that in Rawls' theory, a principle of justice is to serve as part

of a public conception of justice, to which people appeal in adjudicating conflicts over the basic structure of society. A principle of justice is not designed
for an impartial, well-informed god with the power to institute whatever
economic system he decides is most just. Rather, the application of a
principle of justice should be understood by the people affected: whatever
the principle in fact endorses as just ought to be seen as what the principle
endorses by the people whose interests are involved. It is partly for that
reason that the Difference Principle is put in terms of primary social goods
rather than utilities.

Now conflicts between the Difference Principle and the Pareto Principle
arise in cases like that of Figure 2. Suppose that in Figure 2, only social
states w, x, y, and z are feasible. Then only w and y are Pareto efficient.
Consider y: y contains unequal incomes, and is unjust according to the
Difference Principle because a feasible alternative, x, has the same prices and
equal incomes at an intermediate level. x, by a criterion that involves only
incomes, is more just than y. Neither x nor y is a Pareto improvement over
the other, but judged by income, x is the more egalitarian. The same could be
said of w and z.

Should social state z, in which all incomes are equal, be rejected as unjust
because it is Pareto inefficient? Questions of justice aside, everyone prefers
y, with unequal incomes, to z with equal incomes: one group because y shifts
price in a way favorable to them, and the other because they are over-
compensated for an unfavorable price shift. All realize, though, that once
they have shifted to y, there will be a conflict of interest over whether to
shift further to x. They have agreed to resolve such conflicts by looking at
income, and maximizing the minimal income level. By that principle, they
will be committed to the further shift to x. Once people have agreed to make
interpersonal comparisons by an overt standard of income, they should
realize that to accept unequal incomes for the sake of a state everyone prefers
to egalitarian state z may be to raise new questions of justice which, by agreement, will be resolved by moving to a state that differs from z only by a price
shift – a price shift favorable to some and unfavorable to others. Realizing
that, they may find it reasonable to forego a Pareto improvement for the sake
of retaining an overt standard for the resolution of conflicts over the basic
structure of society.

If all this leaves the reader unconvinced, he may wish to adopt a Paretian version of the Difference Principle: that a social state is just iff either it is just according to the old, non-Paretian difference principle, or it is a weak Pareto improvement over some state which is just according to the non-Paretian difference principle. Liberalized slightly more, the principle might read as follows.

Paretian Difference Principle: A social state x is just iff for some feasible social state y,

$$\sim (\exists i)y_i P_i x_i,$$

$$\sim (\exists z, j)(\forall i, k)z_i P_k y_j.$$

This says that a social state is just iff there is a state which is just by the non-Paretian difference principle which no one likes any better.

7. ECONOMICALLY INFLUENCED PREFERENCES

So far, in formulating the difference principle, I have taken preferences as fixed independently of the choice of economic systems. Given the preferences $P_1, \ldots, P_n$ of all n members of society, I have proposed that xBy, individual state x is definitely better than individual state y, iff $(\forall i)xP_i y$. For two social states x and y, I have proposed that $x \mathscr{J} y$ iff for some person, everyone is definitely better off in x than he is in y. What the difference principle now says of basic economic structures is this: in conditions of certainty, one basic economic structure is definitely more just than another iff, where the first would lead to social state x and the second would lead to social state y, we have $x \mathscr{J} y$. A basic economic structure is just iff no alternative basic economic structure is definitely more just than it.

This principle needs to be modified. For preferences among bundles of commodities are clearly not fixed independently of the economic system: alternative basic economic structures will lead to different preferences.[7] Perhaps, then, we should do the following. Let $\mathscr{P}$ be the set of all preference orderings P over bundles of commodities such that for some basic economic structure, if that structure were instituted someone might have preference ordering P. Then for individual states x and y, say xBy iff $(\forall P \in \mathscr{P})xPy$. Alternatively, if we want a more discriminating criterion, let us say that xBy iff $(\exists P \in \mathscr{P})xPy$ and $\sim (\exists P \in \mathscr{P})yPx$. The new B is a strict partial ordering, and so by the Lemma of Section 6, where $x \mathcal{J} y$ means $(\exists j)(\forall i)x_i B y_j$, $\mathcal{J}$ is a partial ordering.

The Difference Principle as now formulated says the following. A basic economic structure is just iff no alternative basic structure is definitely more just. In order to see whether a basic economic structure X is definitely more just than another, Y, do the following: For each basic structure, consider the incomes and prices to which it would lead; let these be given by social states x and y respectively. Consider the situation x_{min} of a worst-off person in x; x_{min} consists of facing the prices of x with the least income of anyone in x. Compare that with y_{min}, the situation of the worst-off person in y. x_{min} is *definitely superior* to y_{min} iff anyone, no matter what basic economic structure had influenced his preferences, would prefer facing situation x_{min} to facing y_{min}. x is *definitely more just* than y iff x_{min} is definitely superior to y_{min}.

This criterion is permissive, in that it allows policies that shift prices so long as anyone prefers the consequent situation of the worst-off – where 'anyone' here includes any sort of person who might emerge from a basic economic structure open for choice at the legislative stage. Suppose, for instance, in an impoverished society, we consider whether to tax everyone to heap lavish subsidies on grand opera. Let X be a basic economic structure, and Y that structure modified by the opera subsidy scheme. Under Y, the poorest people have a slightly lower income than the poorest people have under X, but under Y they have a chance to purchase tickets to grand spectacles that would consume more than their entire income under X. The tickets are expensive, let us suppose, but the poorest person could purchase one if he sacrificed enough in the way of other commodities. In that case, it may be that none of the poorest people do purchase opera tickets, and so by the test of their own preferences, they are worse off with the subsidy scheme than without. It may also be that none of the richer people who do buy tickets would do so if they were as poor as the worst off. None of these facts will settle the issue of whether X is definitely more just than Y, although they establish that Y is not definitely more just than X. The test is rather this: Would any basic economic structure open for choice at the legislative stage produce anyone so devoted to opera that, faced with a choice of being poorest under X and being poorest under Y, he would choose Y, preferring the combination of seeing opera at the cost of an expensive ticket and additional taxes to the alternative of being unable to see it at all. If so, then X is not definitely more just than Y. If moreover, no other basic economic structure open for choice at the legislative stage is definitely more just than either X or Y, then according to the Difference Principle, both X and Y are just. It would then be both just to have the opera subsidy and just not to have it.

Is this criterion overly permissive in the subsidies it allows for the pursuits of the rich – or, for that matter, in the subsidies it allows for the objects of unusual tastes? Perhaps so, but the limits of what it allows should be noted. Call a basic economic structure *eligible* iff it is open for choice at the legislative stage. Which structures are eligible is constrained by feasibility, the priority of liberty, and the priority of having all positions of privilege open to all. Now the set $\mathscr{P}$ consists of those preference orderings that would be formed under any basic economic structure that is eligible. The restriction to eligible structures is crucial. Eligible structures satisfy a constraint of liberty, and so cannot include such things as a compulsory course of brainwashing to determine preferences. The preferences in P, then, do not consist of all preferences that might be produced by a suitable course of conditioning, but simply those preference orderings that would arise under circumstances of liberty given various basic economic structures compatible with liberty. Thus a scheme of opera subsidies may be unjust, since it may be that no one who forms his tastes freely will be so exclusively devoted to opera that even were he among the poorest of the poor, he would willingly sacrifice the price of a ticket and his share of the subsidy in order to see a lavish opera.

8. OTHER PRIMARY SOCIAL GOODS AND UNCERTAIN PROSPECTS

Return now to the problem of primary social goods in general. Just as an income constitutes a choice among alternative bundles of commodities, so does a combination of primary social goods – income and wealth, powers and opportunities – constitute a choice among alternative bundles; now, though, a bundle consists both of commodities and of exercises of powers and opportunities. Call a set of such bundles an *opportunity set*. A rational person certain of his preferences will prefer one opportunity set x to another y iff he prefers the most preferred bundle in x to all bundles in y. We may now treat opportunity sets just as we have been treating individual situations, and formulate a version of the difference principle that applies to primary economic goods in general.

First, though, consider another problem. So far, I have talked as if the choice of a basic economic structure determines, for each person, precisely how he will fare. As Rawls emphasizes, though, such matters are not certain, and the difference principle as Rawls states it evaluates an economic system by the prospects it offers those in the worst starting position (see pp. 78, 96). What is really needed, then, is not an index of incomes or opportunity sets, but an index of prospects over opportunity sets – or, as I have been suggesting, a substitute for an index in the form of a partial ordering.

Even for prospects over income with fixed prices, a ranking by expected money value will not do. For suppose we judge prospects by their expected money payoff. Consider two economic systems: under system X, everyone in a worst-off starting position will get \$ 9000 per year, whereas under system Y, half will get \$ 20,000 and half will starve with no income at all. For those in the worst-off starting positions, then, the expected money value of system Y is higher than that of X. Surely, though, anyone would prefer starting out worst off in system X to starting out worst off in system Y. It would be preposterous to prefer system Y to system X out of a regard for the plight of those in the worst starting positions.

Now the formal method we have been using will apply to a choice among prospects over opportunity sets, and do so without measuring the prospects of those in the worst starting positions by anything like their expected money value. Let $\mathscr{S}$ be the set of all eligible basic economic structures – all basic economic structures open for selection at the legislative stage. Let $\mathscr{X}$ be the set of all prospects over opportunity sets that would be offered anyone under any structure in $\mathscr{S}$, and let variables $\hat{x}, \hat{y}, \hat{z}$ take prospects in $\mathscr{X}$ as values. Let $\mathscr{P}$ be the set of all those preference orderings of members of $\mathscr{X}$ that any one would have some likelihood of developing under any structure in $\mathscr{S}$. Then we can define what it is for one prospect to be definitely better than another:

$$\hat{x}B\hat{y} \quad \text{iff} \quad (\exists P \in \mathscr{P})\hat{x}P\hat{y} \ \& \sim (\exists P \in \mathscr{P})\hat{y}P\hat{x}.$$

Let the *social prospect* offered by a structure $S \in \mathscr{S}$ be the n-tuple $\hat{x} = \langle \hat{x}_1, \ldots, \hat{x}_n \rangle$ of prospects offered by S to the members of the society.

Then where S, $T \in \mathscr{S}$, we say that S is definitely more just than T iff, where $\hat{x}$ is the social prospect offered by S and $\hat{y}$ is the social prospect offered by T,

$$(\exists j)(\forall i)\hat{x}_i B \hat{y}_j.$$

Then basic economic structure S is just iff no alternative structure $T \in \mathscr{S}$ is definitely more just than S. What this says is that a structure is just so long as there is no eligible structure T which improves the prospect of the worst-off in the following sense: that some prospect $\hat{x}$ offered by S is so unappealing compared to all the prospects offered by T that anyone, no matter how his preferences had developed in circumstances of liberty, would prefer each of the prospects offered by T to prospect $\hat{x}$.[8]

I do not know how Rawls would find this as an explication of the difference principle. Clearly it conflicts with the letter of what he says: it evades the construction of an index rather than constructing it, and yields a principle in conflict with the norms of efficiency. It differs from Rawls in its treatment of relevant social positions (pp. 95–100), and avoids aggregation of expectations where Rawls permits it. It may be more permissive than Rawls would like, in that a system is saved from being condemned as less just than an alternative so long as for each prospect it offers, there might be one person whose preferences developed under conditions of liberty and who prefers that prospect to some prospect offered by the alternative.

On the other hand, the principle I have proposed captures a number of aspects of Rawls' theory. It takes seriously the dictum that primary goods are things it is rational to want whatever else one wants, by explaining them as opportunities to choose among bundles of more specific goods. It avoids interpersonal comparisons of satisfaction, and instead looks at prospects for overtly observable income, powers, and opportunities. It avoids the supposition that preferences are fixed independently of the economic structure, and while it respects the constraints of human nature on preferences formed in conditions of liberty, it makes comparisons of prospects independently of the kinds of preferences that would be formed in any particular economic system. Finally, it captures the idea of making the worst off as well off as possible. In a just society, on the explication in this paper, even a person in the worst starting position should realize that, in a sense, it would be impossible to make everyone better off than he in fact is. "In every alternative basic economic structure", he may be told with truth, "there is a starting position that is no better than yours, in the sense that someone with preferences formed under liberty might like that starting position no more than the starting position you actually have".[9]

University of Michigan

NOTES

[1] Page references, unless otherwise indicated, are to Rawls (1971).

[2] Plott (1978) discusses the problem of indexing primary social goods. His approach differs from mine in a number of respects that are discussed in Note 8.

[3] Rawls speaks at one point (p. 97) as if when greater powers and income go together in a society, the indexing problem is avoided. Earlier, he argues only that in that case, the problem is simplified (p. 94).

[4] Rawls speaks not of making the worst off person as well off as possible, but of making the worst-off 'representative man' as well off as possible. He ponders the question of what group's expectations should be aggregated in order to define the worst-off representative man. We are to consider the 'starting places' in society 'properly generalized and aggregated' (p. 96). In this paper I try to avoid aggregation, but in Section 8 I try to do at least part of what Rawls wants to accomplish by aggregation, by taking up his point that we are to apply the maximin criterion to starting positions and the expectations that attach to them, rather than to achieved income levels.

[5] Rawls writes that the difference principle holds among other things "for income and property taxation, for fiscal and economic policy" (1975, p. 97. See also 1977, p. 164.).

[6] Whether a social state is feasible will depend on individuals' preferences, since in a feasible social state, the demand for each commodity must equal its supply. (I owe this observation to Roy Gardner.) Here preferences are taken as fixed; the case of malleable preferences is taken up in Sections 7 and 8.

[7] This problem was brought to my attention by John Bennett. The point is important in Rawls' thought; see 1974, p. 641 and 1975, p. 95.

[8] The approach I have taken to the indexing problem differs from Plott's in a number of respects. Plott considers the problem of indexing bundles of disparate goods, whereas I consider the problem of indexing sets of such bundles, or prospects over sets of such bundles. Plott makes assumptions from which it follows that social states can be weakly ordered by the level of welfare of their worst-off people; I require only a partial ordering. Finally, Plott interprets the difference principle as a requirement that social institutions be designed so as to achieve a most just state whatever well-behaved preferences people may have. I am simply inquiring whether the difference principle can be intelligibly formulated in a way that guarantees that at least one feasible state will be just. Plott shows that the rest of the conditions he imposes are incompatible with Pareto principle, and so in that respect, his conclusion is similar to the conclusion that the weak difference principle is incompatible with the Pareto principle.

[9] I am grateful to John Bennett for extensive and extremely helpful discussion of this paper.

BIBLIOGRAPHY

Plott, Charles R.: 1978, 'Rawls' theory of justice: An impossibility result', in Hans W. Gottinger and Werner Leinfellner (eds.), Decision Theory and Social Ethics: Issues in Social Choice (D. Reidel, Dordrecht, Holland).

Rawls, John: 1971, A Theory of Justice (Harvard University Press, Cambridge, Mass.).

Rawls, John: 1974, 'Reply to Alexander and Musgrave', Quarterly Journal of Economics 88, pp. 633–55.

Rawls, John: 1975, 'A Kantian conception of equality', Cambridge Review, pp. 94–99.

Rawls, John: 1977, 'The basic structure as subject', American Philosophical Quarterly 14, pp. 159–65.

JOHN C. HARSANYI

BAYESIAN DECISION THEORY,
RULE UTILITARIANISM, AND ARROW'S
IMPOSSIBILITY THEOREM

ABSTRACT. The first part of this paper reexamines the logical foundations of Bayesian decision theory and argues that the Bayesian criterion of expected-utility maximization is the *only* decision criterion consistent with rationality. On the other hand, the Bayesian criterion, together with the Pareto optimality requirement, inescapably entails a utilitarian theory of morality. The next sections discuss the role both of cardinal utility and of cardinal interpersonal comparisons of utility in ethics. It is shown that the utilitarian welfare function satisfies all of Arrow's social choice postulates – avoiding the celebrated impossibility theorem by making use of information which is *unavailable* in Arrow's original framework. Finally, rule utilitarianism is contrasted with act utilitarianism and judged to be preferable for the purposes of ethical theory.

1. THE BAYESIAN RATIONALITY POSTULATES

I think it is fair to say[1] that, even today, many social scientists and philosophers fail to realize how strong the arguments really are for the Bayesian theory of rational behavior; and that they have even less appreciation for the important implications that the Bayesian concept of rationality has for ethics and welfare economics. I have recently reexamined the arguments for the Bayesian position, and the moral implications it has (Harsanyi, 1978). Here I will only summarize my main conclusions.

In discussing the concept of rational behavior, I will distinguish behavior under certainty, under risk, and under uncertainty. We act under *certainty* when we can predict the outcome of any specific action we can take. We act under *risk* when we know at least the objective probabilities associated with different possible outcomes. We act under *uncertainty* when even some or all of these probabilities are unknown to us (or are possibly even undefined).

Behavior under certainty amounts to choosing among alternative *situations*, where a situation is characterized by a finite number of economic and noneconomic variables affecting the well-being of the decision maker himself and that of other individuals in the society. Mathematically, a situation A is a point in a finite-dimensional Euclidean space E^m. On the other hand,

Theory and Decision 11 (1979) 289–317.   0040–5833/79/0113–0289$02.90.
Copyright © 1979 by D. Reidel Publishing Co., Dordrecht, Holland, and Boston, U.S.A.

behavior under risk and under uncertainty amount to choosing among alternative *lotteries* whose 'prizes' are situations, i.e., points in E^m. A lottery L can be described as

$$(1.1) \quad L = (A_1 \,|\, e_1 ; \ldots ; A_k \,|\, e_k ; \ldots ; A_K \,|\, e_K),$$

indicating that lottery L will yield situation A_k as prize if event e_k occurs ($k = 1, \ldots, K$). Here, events $e_1, \ldots, e_K$ must be mutually exclusive and exhaustive alternatives. I will call them *conditioning events*. L will be called a *risky* or an *uncertain* lottery according as the decision maker does or does not know the objective probabilities associated with all the conditioning events e_k.

In analyzing the behavior of any individual $i (i = 1, \ldots, n)$, *strict preference* by him will be denoted by $>_i$, *indifference* by $=_i$, and *nonstrict preference* by $\geq_i$. Rational behavior by this individual i under certainty can be characterized by two postulates.

Postulate 1: Complete preordering. The nonstrict preferences of individual i establish a complete preordering over the space E^m of all possible situations.

Postulate 2: Continuity. Suppose that some sequence $A_1, A_2, \ldots,$ of situations converges to a given situation A_0, and that another sequence $B_1, B_2, \ldots,$ of situations converges to B_0, with $A_k \geq_i B_k$ for $k = 1, 2, \ldots$. Then, $A_0 \geq_i B_0$.

For convenience, I will call these two postulates the *basic utility axioms*. They can be used to establish the following theorem.

Theorem 1: Utility maximization. If an individual's preferences satisfy the two basic utility axioms then his behavior will be equivalent to maximizing a well-defined (ordinal or cardinal) utility function U_i. (See Debreu, 1959, pp. 55–59.)

To characterize rational behavior under risk and uncertainty, we need two additional rationality postulates.

Postulate 3: Probabilistic equivalence. Let

$$(1.2) \quad L = (A_1 \,|\, e_1 ; \ldots ; A_K \,|\, e_K) \quad \text{and} \quad L^* = (A_1 \,|\, e_1^* ; \ldots ; A_K \,|\, e_K^*),$$

and suppose that the decision maker knows the objective probabilities associated with events $e_1, \ldots, e_K$ and with events $e_1^*, \ldots, e_K^*$, and knows that these probabilities satisfy

$$(1.3) \qquad \mathrm{Prob}\,(e_k) = \mathrm{Prob}\,(e_k^*) \qquad \text{for } k = 1, \ldots, K.$$

Then, he will be indifferent between lotteries L and L^*.

In other words, a rational individual will be indifferent between two risky lotteries yielding him the same prizes with the same probabilities – even if the two lotteries use quite different physical processes to generate these probabilities. (In particular, he will be indifferent between a one-stage and a two-stage lottery if they are probabilistically equivalent. Thus, Postulate 3 implies von Neumann and Morgenstern's postulate on compound lotteries.)

Postulate 4: The sure-thing principle. Suppose that $A_k^* \geqq_i A_k$ for $k = 1, \ldots, K$. Then

$$(1.4) \qquad (A_1^* | e_1; \ldots; A_K^* | e_K) \geqq_i (A_1 | e_1; \ldots; A_K | e_K).$$

In other words, other things being equal, a rational individual will not prefer a lottery yielding less desirable prizes over a lottery yielding more desirable prizes.

In my paper already quoted, I tried to show that

(A) Postulates 3 and 4 are extremely compelling rationality requirements, but that this statement is subject to two conditions:

(A*) The utility $U_i(A_k)$ of any prize A_k must be *independent* of its conditioning event e_k.

(A**) The decision maker must take a strictly *result-oriented* attitude toward lotteries, in the sense of deriving all his utility and disutility from the prizes he may or may not win, rather than from the act of gambling itself; i.e., from the nervous tension associated with gambling (assumption of no direct utility or disutility for gambling).

I have also tried to show that – if the 'prizes' of the lotteries are properly defined then – conditions (A*) and (A**) and, therefore, also Postulates 3 and 4 will always be satisfied for serious *policy* decisions and for personal *moral* decisions – but that, in the case of gambling done for *entertainment*, Postulate 3 will often be violated because condition (A**) will fail.

I will describe Postulates 1, 2, 3, and 4 together as the *Bayesian rationality postulates*. They can be used to establish the following theorem.

Theorem 2: Expected utility maximization. If an individual's preferences satisfy the four Bayesian rationality postulates, then he will have a (cardinal) utility function U_i which assigns, to any lottery L of form (1.1), a utility equal to its expected utility, that is, equal to the quantity

$$(1.5) \quad U_i(L) = \sum_{k=1}^{K} p_k U_i(A_k);$$

and this is the quantity he will maximize by his behavior. Here p_k is the probability associated with event e_k ($k = 1, \ldots, K$). More specifically, if L is a risky lottery, then p_k must be interpreted as the objective probability $p_k = \text{Prob}(e_k)$ of this event e_k; whereas if L is an uncertain lottery, then p_k must be interpreted as the subjective probability $p_k = \text{Prob}_i^*(e_k)$ that the decision-maker, individual i, chooses to assign to e_k.

Property (1.5) is called the *expected-utility* property, and any utility function U_i possessing this property is called a *von Neumann-Morgenstern utility function*. (For risky lotteries, a proof of Theorem 2 can be found in von Neumann and Morgenstern, 1947, pp. 617–628. For uncertain lotteries, a proof can be obtained by a slight modification of that given in Anscombe and Aumann, 1963. See Harsanyi, 1978.)

Thus, we can conclude that the Bayesian rationality postulates and, therefore, the requirement of expected-utility maximization, are absolutely compelling requirements for rationality – at least in the cases of serious *policy* decisions and of personal *moral* decisions.

2. THE LINEARITY OF SOCIAL WELFARE FUNCTIONS

In a much earlier paper (Harsanyi, 1955, pp. 313–14), I have established the following result. Let us describe the preferences guiding an individual's every-day behavior, and expressed by his utility function U_i, as his *personal* preferences. Most people's personal preferences are not completely selfish, yet usually no doubt give greater weight to their own interests and to the interests of their closest associates than they give to the interests of complete strangers. But people's behavior is not always guided by these more or less self-serving personal preferences: sometimes it is guided by much more impartial and impersonal considerations. This (one hopes) will often be the case when they act as judges or as public officials of other kinds. More importantly from our present point of view, this will be the case when they make *moral value judgments*. (Of course, people often make value judgments based on self-interest or on a partial concern for the interests of their family and friends. But such judgments are mere *judgments of personal preference* and do not qualify as genuine moral value judgments.)

As distinguished from an individual's *personal* preferences as defined above, the impartial and impersonal considerations he follows in his moral value judgments (and in his moral decisions) will be called his *moral* preferences. His personal preferences are his preferences as they normally are; while his moral preferences are those preferences that he exhibits in those – possibly quite rare – moments when he forces a special impartial and impersonal attitude, i.e., a *moral* attitude, upon himself. Whereas the personal preferences of a given individual i are expressed by his utility function U_i, his moral preferences are expressed by his *social welfare function* W_i.

Let me now consider a society consisting of n individuals. Suppose we analyze the moral preferences of one particular individual, to be called individual j (who normally will be but need not be himself a member of the society under consideration). I propose the following three axioms.

Axiom A: Individual rationality. The personal preferences of *all* n individuals satisfy the four Bayesian rationality postulates.

Axiom B: Rationality of moral preferences. The moral preferences of individual j satisfy the four Bayesian rationality postulates.

Axiom C: Pareto optimality. Suppose that at least *one* of the n individuals personally prefers social situation A over social situation B, and that none of the other individuals prefers B over A. Then, individual j will morally prefer A over B.

Axiom A is an obvious rationality requirement. So is Axiom B: when acting in the public interest or when making moral value judgments we must follow at least as high standards of rationality as we follow in pursuing our personal interests. Thus, if we are to follow the Bayesian rationality postulates in our personal affairs as Axiom A suggests, then we are under an even stronger obligation to follow these postulates in public affairs or in making moral value judgments. Finally, Axiom C of course is not a rationality postulate but rather is a moral principle – but it is surely a rather noncontroversial moral principle.

The three axioms imply the following theorem.

Theorem 3: Linearity of the social welfare function. The social welfare function W_j of individual j must be a real-valued function over all social situations A, and must have the mathematical form

$$(2.1) \quad W_j(A) = \sum_{i=1}^{n} \alpha_{ij} U_i(A) \quad \text{with} \quad \alpha_{ij} > 0 \quad \text{for} \quad i = 1, \ldots, n.$$

Thus, the Bayesian rationality postulates, together with a Pareto-optimality requirement, logically *entail* a utilitarian social welfare function. Note that the theorem *does not depend on the possibility of interpersonal utility comparisons* – though if such utility comparisons are not admitted then the coefficients $\alpha_{1j}, \ldots, \alpha_{nj}$ will have to be based completely on individual j's – possibly quite subjective – personal value judgments. On the other hand, if interpersonal comparisons of utilities (of utility differences) are admitted, then we can add a fourth axiom.

Axiom D: Equal treatment of all individuals. If all individuals' utility functions $U_1, \ldots, U_n$ are expressed in *equal utility units* (as judged by individual j on the basis of interpersonal utility comparisons), then the social welfare function W_j of individual j must assign the *same* weight to all these utility functions.

Given this axiom, we can conclude that the coefficients α_{ij} in (2.1) will satisfy:

$$(2.2) \quad \alpha_{1j} = \cdots = \alpha_{nj} = \alpha.$$

3 . THE EQUIPROBABILITY MODEL
FOR MORAL VALUE JUDGMENTS

Instead of the axiomatic approach used in Section 2, we can establish a similar (and, indeed, a slightly stronger) result by means of a constructive model. Suppose that individual j wants to make a moral value judgment as to the relative merits of two social situations A and B. I have argued that his judgment will be a true moral value judgment only if it is based on impartial and impersonal criteria. But I have not given any formal definition for this impartiality and impersonality requirement. I now propose to do so.

One way of assuring that individual j's judgment would not be unduly influenced by his personal self-interest would be to require that he should choose between situations A and B *without knowing* what his personal social position would be under either situation – more specifically, to make him choose between them on the assumption that he would have the *same probability* $1/n$ of ending up in any one of the available n social positions. This assumption I will call the *equiprobability model* of moral value judgments. By Theorem 2, under these conditions, individual j – if he follows the Bayesian standards of rationality – will make his choice so as to maximize his expected utility: which in this case will amount to choosing that particular situation that corresponds to the *highest average utility level* in the society.[2]

Of course, in order for him to make an impartial and impersonal moral value judgment, it is not really necessary that the decision maker literally should not *know* what his personal position would be in the two social situations. It is quite sufficient if he makes a serious effort to *disregard* this morally irrelevant piece of information when he is making his moral value judgment. Our analysis can be summarized in

Theorem 4: The social welfare function as arithmetic mean of individual utilities. Under the equiprobability model, an individual following the Bayesian rationality postulates will make his moral value judgments so as to maximize the social welfare function

$$(3.1) \quad W_j(A) = \frac{1}{n} \sum_{i=1}^{n} U_i(A).$$

Once more, our theorem yields a social welfare function in keeping with the utilitarian tradition. Note that, unlike Theorem 3, Theorem 4, and the equiprobability model itself, *do* presuppose the possibility of interpersonal comparison of utility differences.

4. CARDINAL UTILITY

The mathematical facts stated in my Theorems 1 to 4 are, of course, non-controversial. The Pareto-optimality requirement used in Theorem 3 is widely accepted. The four Bayesian rationality postulates I have used also enjoy fairly wide acceptance now.[3] Nevertheless, until quite recently, there was considerable resistance to utilitarian theory. Historically, this resistance had two main sources: one was opposition to the very concept of *cardinal utility*; the other was a rejection of *interpersonal utility comparisons*. But in the last few years several distinguished economists have come out in favor of utilitarianism or related ideas.

Thus, Henri Theil endorsed my version of utilitarian theory in 1964 (Theil, 1964, p. 336). Paul Samuelson, who used to be a strict ordinalist, announced his support in 1974, calling my theory 'one of the few quantum jumps' in welfare economics (Samuelson, 1974, pp. 1266–67, footnote). Even Ken Arrow, who clearly remains an ordinalist, has recently argued, with some reservations, in favor of interpersonal utility comparisons, even if only within the ordinalist framework (Arrow, 1977; cf. also Arrow, 1963, pp. 114–15).

Let me start my discussion of cardinal utility vs. ordinal utility with the relevant definitions. A utility function U_i of a given individual i is an *ordinal* utility function if it is meant to compare at least *utility levels*, in the sense of telling us, for any pair of alternatives A and B, whether $U_i(A) >$ or $<$ or $= U_i(B)$. *Any* well-defined utility function must be at least ordinal. The basic measurement operation underlying ordinal utility functions U_i is to observe whether individual i in actual choice situations prefers A to B, or prefers B to A, or is indifferent between the two.

A utility function U_i is a *cardinal* utility function if it is meant to compare, not only utility levels, but also *utility differences*, in the sense of telling us, for any triplet of situations A, B, and C, whether $\Delta U_i(A, B) = U_i(A) - U_i(B) >$ or $<$ or $= \Delta U_i(B, C) = U_i(B) - U_i(C)$. Observation of individual i's behavior under *certainty* in general does not provide any measurement operation that could define a cardinal utility function.[4] But observation of his behavior under *risk* (and under uncertainty) does – if his behavior is consistent with the Bayesian rationality postulates.

Let $(X, \frac{1}{2}; Y, \frac{1}{2})$ denote a risky lottery yielding the two alternative prizes X and Y with the same probability $\frac{1}{2}$. Then, we can define the required measurement operation as follows. Observe whether individual i prefers lottery $L = (A, \frac{1}{2}; C, \frac{1}{2})$ over the certainty of obtaining prize B or not. If he does then, in view of Theorem 2, we can infer that

$$(4.1) \quad \tfrac{1}{2} U_i(A) + \tfrac{1}{2} U_i(C) > U_i(B),$$

which is algebraically equivalent to

$$(4.2) \quad \Delta U_i(A, B) = U_i(A) - U_i(B) > \Delta U_i(B, C) = U_i(B) - U_i(C).$$

On the other hand, if he prefers B over L (or is indifferent between them) then, by similar reasoning, we can infer that $\Delta U_i(A, B) < [\text{or} =] U_i(B, C)$.

We can put the distinction between ordinal utility and cardinal utility into a broader perspective by noting that similar distinctions arise in other disciplines. In particular, what economists like to describe as a distinction between *ordinal* and *cardinal* quantities, is discussed in the psychological literature, with considerable sophistication, as a distinction between *ordinal* scales and *interval* scales.

Thus, we can conclude that Bayesian theory provides an operationally meaningful measurement operation for defining cardinal utility functions (von Neumann-Morgenstern utility functions). While *before* the emergence of Bayesian theory economists had very good reasons for rejecting cardinal utility functions, such an attitude now lacks any justification whatever *after* the emergence of Bayesian theory. Even less justification is there for trying to evade the issue by means of terminological dodges, such as calling cardinal utility functions 'measurable ordinal utility functions' or 'ordinal utility functions defined over lotteries', etc., as some economists have proposed to do.

Of course, all cardinal utility functions are also ordinal (because they do always permit comparisons *both* of utility levels *and* of utility differences); and they are 'measurable' and are in fact 'defined over lotteries'. But they are not *merely* ordinal (i.e., noncardinal) utility functions. We can say the same thing about verbal evasions of this type as Bertrand Russell once said in another connection: Such terminological moves have "all the advantages of theft over honest toil". What they amount to is making full use of the superior analytic power of cardinal utility theory while pretending to be true believers of an ordinalist orthodoxy.

Once we have defined cardinal utility functions for the various economic agents on the basis of their behavior under *risk* and *uncertainty*, these cardinal utility functions become available also for analysing their behavior under *certainty*. Many economists fail to realize how much simplification, and how much additional analytical power, these cardinal utility functions provide for economic theory (especially for the theory of consumer behavior). For example, Hicks has devoted a good deal of effort to finding purely ordinal definitions for *substitution* and for *complementarity* between commodities (Hicks, 1939, Chap. III). Though he did come up with very ingenious definitions, these are, inevitably, quite cumbersome and complicated. But, once we are free to use cardinal utility functions, substitution and complimentarity once more become very simple and transparent concepts, which even freshmen can easily understand.

Let $U_i(A)$, $U_i(B)$, and $U_i(A \& B)$, denote the cardinal utility that individual i derives from (1) consuming commodity A (without consuming commodity B), (2) consuming B (without A), and (3) consuming commodities A and B *together*, respectively. Then, we can define A and B to be *substitutes*, or to be *complements*, or to be *independent* commodities, according as $U_i(A \& B)$ is smaller than, larger than, or equal to, the sum $U_i(A) + U_i(B)$. Reflection will show that these definitions fully express the intuitive meanings of substitution, complementarity, and independence.

The concepts of substitutes and of complements can also be used to explain why a person's von Neumann-Morgenstern (vNM) cardinal utility function may show both ranges of *decreasing* and of *increasing* marginal utility. As is well known, most commodities are (usually weak) substitutes for each other. In ranges where this relationship of substitution predominates among the commodities purchased by the consumer, his cardinal utility function will tend to show *decreasing* marginal utility; whereas in those – much less common – ranges where the relationship of complementarity predominates, there will be a tendency to *increasing* marginal utility.

It is somewhat surprising that, even among economists who fully accept a use of vNM cardinal utility functions in the theory of risk-taking (and possibly also in other fields of positive economics), there is some opposition to a use of such functions in ethics and welfare economics. It is argued that vNM utility functions merely express people's 'tastes . . . for gambling', which have no ethical significance (e.g., Arrow, 1963, p. 10). But do vNM utility functions in fact merely express people's tastes for gambling?

These utility functions certainly do not express people's tastes for gambling in the sense of indicating their direct utilities or disutilities for gambling. Indeed, Condition (A**) of Section 1 specifically *excludes* any such direct utilities or disutilities from people's vNM utility functions. According to Condition (A**), these vNM utility functions express people's tastes for gambling only in the sense of indicating their strictly *result-oriented* preferences among alternative lotteries. In other words, they express people's preferences among alternative lotteries as determined by the *relative importance* (i.e., the cardinal utility) these people assign to the various *prizes* offered by these lotteries.

For example, suppose that individual i is willing to pay \$5 for a lottery ticket that will yield him a $1/1000$ chance of winning \$1000, and suppose that he has a strictly result-oriented attitude toward this lottery. How will the theory of vNM utility functions explain the fact that he is willing to gamble at such highly unfavorable odds? The explanation will be obviously in terms of the *relative importance* he assigns to the possibility of winning \$1000 as against the relative importance he assigns to the possibility of losing the \$5 he will invest. The former can be measured by the cardinal utility $U_i(1000)$ he assigns to \$1000 while the latter can be measured by the cardinal utility $U_i(5)$ he assigns to \$5. His gambling behavior can be explained by the fact that $(1/1000)U_i(1000) \geq U_i(5)$, i.e., by the fact that the ratio of $U_i(1000)$ to $U_i(5)$ is at least $1000 : 1$ even though the ratio of the corresponding money prizes is only $1000 : 5 = 200 : 1$.

It is easy to verify that, if individual i assigns such a high cardinal utility to winning \$1000 (as compared with the cardinal disutility he assigns to losing \$5), then his vNM utility function must show *increasing* marginal utility over (part or whole) of the relevant range. As I have argued above, such a range of increasing marginal utility will arise when there are important *complementarities* among the commodities he would buy at these income levels. For example, he may want to use the \$1000 for buying (say) various complementary pieces of expensive ski equipment. Very often this complementarity will arise from a desire to buy a large *indivisible* commodity unit. (Different parts or components of such a unit are complementary goods from the consumer's point of view.) For instance, the great importance he attaches to obtaining \$1000 will become readily understandable if he wants to use the money as deposit on a very badly needed car, etc.

Thus, in fact, a vNM utility function does much more than express a given individual's taste or distaste for gambling. Rather, it uses this individual's preferences among lotteries to measure the relative importance (i.e., the cardinal utility) he assigns to various possible prizes, and by this means to measure the relative intensities of his various wants.

Accordingly, vNM utility functions have a natural place in ethics. Their use enables us to construct social welfare functions reflecting, not only people's actual wants but also the *relative importance* these people themselves assign to their various wants; and reflecting, not only people's actual preferences but also the *relative intensities* of these preferences. The use of vNM utility functions implies that, other things being equal, our social welfare function will assign higher social priorities to those objectives to which the individual members of society themselves assign high personal priorities – as shown by their willingness to take considerable *risks*, if the need arises, for attaining these objectives.

5. INTERPERSONAL UTILITY COMPARISONS

We all make, or at least attempt to make, interpersonal utility comparisons all the time. We have to decide again and again which member of our family, which friend of ours, or which charitable cause, should be given some specified part of our time, or of our money, or of our other scarce resources, whether in the form of a special favor, a small present, or even a large donation, etc. In making such a decision, one of the considerations will always be who needs our help more or who can derive a greater benefit from it, which by necessity amounts to making interpersonal utility comparisons.

It is not hard to see what the psychological basis is for such interpersonal comparisons. They are based on *imaginative empathy*, on an ability to imagine ourselves to be in the shoes of other people. Of course, with the possible exception of some exceedingly unsophisticated individuals, all of us will surely realize that we cannot compare other people's utility levels with our own by merely considering the differences between their objective conditions (their wealth, social position, health, etc.) and ours; rather, we have to make proper allowances also for possible differences between these people's tastes or personal attitudes and our own. (Having to eat fish every day would make me rather unhappy because I am not very fond of seafood;

but it would be very foolish for me to overlook the fact that other people might greatly enjoy eating a lot of fish.) As G. B. Shaw once said, "Do not do unto others what you want them to do unto you; for they might have different tastes".

Thus, if I try to estimate another person's present utility level, then I cannot simply ask how much utility *I* would derive from the objective situation he is in, with *my* own taste and my own personal attitudes. Rather, I have to ask how much utility *I* would derive from the objective situation he is in if I had *his* taste and his personal attitudes – and, more fundamentally, if my personality had been formed by those biological, psychological, sociological, and cultural forces that had formed *his* personality. Such questions are often very hard to answer but they are perfectly meaningful questions. In principle at least, they can be answered by the usual methods of empirical science, though in practice we often cannot do any better than use our own best judgment in giving highly tentative answers to them.

Of course, if interpersonal utility comparisons are to be used in any formal theory then they must satisfy certain *consistency requirements* (consistency axioms). I have described the consistency requirements we need elsewhere (Harsanyi, 1977a, pp. 51–60).

From a logical point of view, interpersonal utility comparisons are based on what I have called the *similarity postulate* (Harsanyi, 1977c, pp. 638–642). By this I mean the principle that, given the basic similarity in human nature (i.e., in the fundamental psychological laws governing human behavior and human attitudes), it is reasonable to assume that different people will show very *similar* psychological reactions to any given objective situation, and will derive much the *same* utility or disutility from it – *once proper allowances have been made for any empirically observed differences* in their biological make-ups, in their social positions, in their educational and cultural backgrounds and, more generally, in their past life histories. In other words, in the absence of clear evidence to the contrary, the presumption must always be that people's behavior and psychological reactions will be similar in similar situations.

Thus, from a philosophical point of view, interpersonal utility comparisons have the status of *empirical* hypotheses based on a *nonempirical* a priori principle, the similarity postulate. Of course, as philosophers of science have often pointed out, our choice of empirical hypotheses is *always* guided by

various nonempirical a priori criteria, such as simplicity, parsimony, a preference for the 'least arbitrary' hypotheses, etc. Our similarity postulate is a
nonempirical principle of the same general type.

The similarity postulate is the logical basis, not only for interpersonal comparisons of utility, but also *for assigning other people conscious experiences
at all*. From a strictly empirical point of view, a world in which I were the
only person with conscious experiences and in which all other people were
unfeeling robots would be industinguishable from our actual world, where, as
common sense tells us, *all* normal humans are fully self-conscious human
beings. If those people who rejected the similarity postulate, and rejected
interpersonal utility comparisons based on this postulate, were really consistent then they would have to conclude that they were the only persons
with a conscious awareness, and that all other people were mindless automata.

One must really wonder how anybody could ever have come to reject
interpersonal utility comparisons. The answer is no doubt historical: it was
largely a result of uncritical acceptance of a seriously mistaken – and by now
completely superseded – philosophical doctrine, that of logical positivism.

Let me make it clear that I have the strongest admiration for the logical
positivists, even though I disagree with them on many points. We are no
doubt greatly indebted to them intellectually for their firm insistence on
greater logical and mathematical rigor in philosophic and in scientific arguments, and for their close scrutiny of all philosophic and scientific theories
for actual empirical content. But at the same time, it is now perfectly clear
that the logical positivists, especially in their early period, followed an unduly
rigid form of empiricism, and had very insufficient appreciation for the
crucial role that *nonempirical* a priori principles, such as the similarity postulate, actually play in the empirical sciences and in philosophy.

It seems to me that, after so many years, now the time has come both
for economists and for philosophers to have a fresh look at the whole
question of interpersonal utility comparisons, without being sidetracked by
thoroughly outmoded philosophical prejudices.

6. UTILITARIAN SOCIAL WELFARE FUNCTIONS AND ARROW'S IMPOSSIBILITY THEOREM

Arrow (1963) has proposed five 'Conditions' (axioms) that any social welfare function ought to satisfy and then showed that, in actual fact, no social welfare function can possibly satisfy all of these Conditions. Arrow's entire analysis has been stated within a strictly *ordinal* framework. The question arises whether Arrow's negative conclusions retain their validity also when use of *cardinal* utility functions is admitted.

In fact, it turns out that use of cardinal utilities is insufficient to enable us to avoid Arrow's Impossibility Theorem. On the other hand, I propose to show that we can completely escape (the cardinal counterpart of) Arrow's Theorem if *both* cardinal utility functions *and* interpersonal utility comparisons are admitted, as is the case under the utilitarian conceptual framework.

I will first restate Arrow's five Conditions in a way appropriate for a model admitting cardinal utility functions. Under such a model, the dependence of (individual *j*'s) social welfare function on the individual cardinal utility functions $U_1, \ldots, U_n$ can be expressed by writing, for any alternative A, an equation of the form

$$(6.1) \qquad W_j(A) = F_j[U_1(A), \ldots, U_n(A)].$$

Here W_j, as well as $U_1, \ldots, U_n$, are functions from the set of all possible social situations (i.e., from the set of all possible alternatives) to the set of real numbers. On the other hand, F_j is a function from the set of all real *n*-vectors (representing the various individuals' utilities for given social situations) to the set of real numbers. Equivalently, from another point of view, F_j is a mapping from the set of all possible *n*-tuples of utility functions to the set of all possible social welfare functions.

Since I am using the term '*social welfare function*' to describe the function W_j, I will describe F_j as the *social welfare mapping*. The purpose of the following five Conditions is to characterize this mapping F_j.

Condition 1: Unrestricted domain. The mapping F_j is defined for all possible *n*-tuples of cardinal utility functions $U_1, \ldots, U_n$.

Condition 2: Positive association of social and individual values. (Pareto optimality.) For any pair of alternatives A and B, the mapping F_j makes the social welfare difference $\Delta W_j(A, B) = W_j(A) - W_j(B)$ a strictly increasing function of the individual utility differences $\Delta U_i(A, B) = U_i(A) - U_i(B)$ for $i = 1, \ldots, n$.

Condition 3: Independence of irrelevant alternatives. For any alternative A, the mapping F_j makes the social welfare value $W_j(A)$ of A depend *only* on the individual utilities $U_i(A)$ associated with A, and makes $W_j(A)$ independent of the individual utilities $U_i(B)$ associated with *other* alternatives $B \neq A$, and of *any* other variables whatever (such as characteristics of the feasible set, other environmental variables, etc.).

Condition 4: Nonimposition. For any pair of alternatives A and B, there is at least one n-tuple of individual utility functions $U_1, \ldots, U_n$ yielding

$$(6.2) \qquad W_j(A) = F_j[U_1(A), \ldots, U_n(A)] > W_j(B)$$
$$= F_j[U_1(B), \ldots, U_n(B)] ;$$

there is at least one other n-tuple of such functions yielding the opposite inequality; and there is at least one n-tuple of such functions satisfying (6.2) with an equality sign.

Condition 5: Nondictatorship. There is no individual i with so strong an influence on the social welfare function W_j that $U_i(A) > U_i(B)$ would always imply $W_j(A) > W_j(B)$, regardless of the utilities $U_k(A)$ and $U_k(B)$ that the *other* individuals $k \neq i$ associate with A and with B.

I now propose to state the following theorem.

Theorem 5: Utilitarianism and Arrow's Impossibility Theorem. Given the admissibility of interpersonal utility comparisons, the social welfare function W_j defined by Theorem 4 – as well as that defined by Theorem 3 and Equation (2.2) – satisfy Conditions 1 to 5 as stated above.

Proof. To save space, I will state the proof of the theorem only for the social welfare function W_j defined by Theorem 4. (The proof for the social welfare function defined by Theorem 3 is, however, very similar.) W_j satisfies Condition 1 since *any* n-tuple of cardinal utility functions $U_1, \ldots, U_n$ can be used in Equation (3.1). It satisfies Condition 2 since (3.1) makes the quantity $\Delta W_j(A, B) = W_j(A) - W_j(B)$ a linear combination, with strictly positive coefficients, of the quantities $\Delta U_i(A, B) = U_i(A) - U_i(B)$. That W_j satisfies Condition 3 directly follows from (3.1). As to Condition 4, it is easy to give an example for an n-tuple $(U_1, \ldots, U_n)$ yielding a function W_j satisfying (6.2): for example, this will be the case if all differences $\Delta U_i(A, B) = U_i(A) - U_i(B)$ are positive. On the other hand, if all these differences are negative, then we obtain a function W_j satisfying the reverse of Inequality (6.2). Finally, if all these differences are zero, then we obtain a function W_j satisfying (6.2) with an equality sign. Moreover, W_j satisfies Condition 5: Even if, for a given individual i, the difference $\Delta U_i(A, B)$ is *positive* and is very, very large, the difference $\Delta W_j(A, B)$ will still be *negative* in case the differences $\Delta U_k(A, B)$, $k \neq i$, are negative for enough individuals k, and are large enough in their absolute values. Thus, no individual i will be a dictator.

In view of Theorem 5, we can infer that the cardinal counterpart of Arrow's Impossibility Theorem is *false*. Obviously, this must mean that the cardinal analogue of Arrow's proof does not go through. In fact, it is easy to see where the proof gets stuck in the cardinal case.

A crucial step in Arrow's proof is to construct some subset V of the n individuals with the following properties:

1. All individuals i in set V prefer some alternative A over some alternative B.

2. The social ordering, likewise, prefers A over B.

3. Yet, all individuals i' not in V (if there are any such individuals) have opposite preferences, preferring B over A.

Then, one shows that, by using Arrow's axioms, one can infer that set V is a *decisive set* for A against B, in the sense that, as long as all individuals i in V prefer A over B, the social ordering must do the same, *regardless* of what the preferences of all individuals not in V may be.

The cardinal-utility analogue of this argument would be this. Suppose that:

1^*. For all individuals i in set V, we have

$$\Delta U_i(A, B) = U_i(A) - U_i(B) > 0.$$

2^*. For the social welfare function W_j, we likewise have

$$\Delta W_j(A, B) = W_j(A) - W_j(B) > 0.$$

3^*. Yet, for all individuals i' not in V, we have

$$\Delta U_{i'}(A, B) = U_{i'}(A) - U_{i'}(B) < 0.$$

Now, try to infer from this the conclusion that V is a *decisive set* for A against B in the sense that, if statement 1* is satisfied, then statement 2* will always follow, regardless of the signs of the utility differences $\Delta U_{i'}(A, B)$ for the individuals not in set V. This conclusion will *fail* for the following reason.

There is a crucial difference between the ordinal case and the cardinal case. In the *ordinal* case (if we represent the individual preferences and the social ordering by numerical functions) the sign of the social welfare difference $W_j(A, B)$ will depend only on the *signs* of the individual utility differences $U_i(A, B)$ and $U_{i'}(A, B)$. In contrast, in the *cardinal* case, the sign of $W_j(A, B)$ will depend, not only on the *signs* of these individual utility differences, but also on their actual *magnitudes* (absolute values). In other words, in the cardinal case, the direction of social preference will depend, not only on the *directions* of the individual preferences, but also on the *intensities* of these preferences.

For example, take a society consisting of three individuals. Suppose that 1 and 2 prefer A over B whereas 3 prefers B over A. Thus, we can take $V = \{1, 2\}$. More particularly, suppose that

$$(6.3) \quad U_1(A) - U_1(B) = U_2(A) - U_2(B) = 18 > 0,$$

whereas

$$(6.4) \quad U_3(A) - U_3(B) = -18 < 0.$$

Then, in view of 3.1, we will have

$$(6.5) \quad W_j(A) - W_j(B) = (18 + 18 - 18)/3 = 6 > 0.$$

Yet, we cannot infer from this that set V is a *decisive set* for A against B. For, let us now suppose that

$$(6.6) \quad U_1(A) - U_1(B) = U_2(A) - U_2(B) = 6 > 0,$$

whereas

$$(6.7) \quad U_3(A) - U_3(B) = -18 < 0, \text{ as before.}$$

Then, in view of 3.1, we will have

$$(6.8) \quad W_j(A) - W_j(B) = (6 + 6 - 18)/3 = -2 < 0.$$

Thus, now the social ordering will prefer B over A even though both members of set V still prefer A over B. Though all utility differences still have the same signs as before, the preferences of individuals 1 and 2 have become less intense, and this fact will be sufficient to reverse the direction of social ordering between A and B.

Let me add that, under certain conditions, Theorem 5 can be extended also to social welfare functions W_j defined by Theorem 3 *without* making use of interpersonal utility comparisons (i.e., without making use of Equation (2.2)): whether this can be done or not depends on the actual way individual j chooses the coefficients α_{ij} of Equation (2.1).[5]

Theorem 5 shows that the social welfare functions defined by Theorems 3 and 4 fully escape the difficulties posed by Arrow's Impossibility Theorem (at least if interpersonal comparisons are admitted): these social welfare functions do exist, even though they satisfy all of Arrow's Conditions as restated for the cardinal case. There are statements in the literature that *no* social welfare function can escape the problems raised by Arrow's theorem: these statements are evidently wrong.

How do my social welfare functions manage to evade the theorem? The answer is that my theory makes use of two pieces of information that are unavailable within Arrow's conceptual framework:

(1) It makes use of *cardinal* individual utility functions U_i.

(2) It makes use of *interpersonal utility comparisons* (or some functionally equivalent alternative measuring operation if interpersonal comparisons are not admitted).

Property (1) is essential because *ordinal* utility functions are susceptible of arbitrary *monotone* transformations; and such transformations would destroy Equation (2.1) (with $\alpha_{ij} = \alpha$ for $i = 1, \ldots, n$) and Equation (3.1). Even property (1) without property (2) would be insufficient because independently defined cardinal utility functions are still susceptible of arbitrary *linear* monotone transformations; and such transformations would just as necessarily destroy the relevant equations. (For example, either equation will lose its validity if we multiply U_1 by 2 while keeping all other utility functions U_i with $i \neq 1$ unchanged.) But, once we have *both* properties, we can require that all utility functions U_i should be expressed in *equal* utility units (as judged by individual j on the basis of interpersonal utility comparisons); and we can restrict the set of permissible transformations to those linear monotone transformations which *maintain* this equality among the different individuals' utility units, i.e., which multiply all individuals' utility functions by the *same* positive constant. Such transformations will preserve the validity of Equations (2.1) and (3.1) (if then W_j is also multiplied by the same positive constant).[6]

Several authors have noted that the social welfare functions defined by my theory escape Arrow's Impossibility Theorem. Sen (1970, pp. 128–130) has also correctly pointed out that the reason lies in my assuming *both* cardinality and comparability, and that cardinality alone would not do. Other authors, however, have incorrectly claimed that my social welfare functions avoid Arrow's theorem because they are always defined only for *one* particular n-tuple $(U_1, \ldots, U_n)$ of individual utility functions, and are not based on any *general* mapping from *all* possible n-tuples of utility functions to some appropriate social welfare functions W_j (see, e.g., Kalai and Schmeidler, 1977, pp. 1431–32). But this claim is mistaken because Equation (3.1) – or Equations (2.1) and (2.2) – do define such a general mapping (which we have called the mapping F_j above): they do permit us to construct a social welfare function W_j for *every* possible n-tuple $(U_1, \ldots, U_n)$ of individual cardinal utility functions.[7]

7. SOCIAL DECISIONS AND INDIVIDUAL MORAL DECISIONS

A study of possible social welfare functions is usually undertaken for two main purposes: to gain a better understanding of *social decision-making processes* (such as elections, committee decisions, administrative decisions, the market mechanism, etc.) or to gain a better understanding of individual *moral decisions*. These two types of decision processes do essentially coincide in one very special case, viz., when society delegates to a public official the task of making a social decision in accordance with his own best judgment, and in accordance with the dictates of his own moral conscience. But in all other cases, a study of social decisions, made by a number (sometimes by many millions) of individuals, is a very different problem from a study of moral decisions, made by one individual – even if this one individual makes his decision partly or wholly in terms of *social* criteria.

To be sure, a normative study of social decisions does have a *moral* aspect. A moral philosopher may very appropriately ask the question, What are the moral obligations of the various participants in a social decision process? What are the moral criteria that the individual voter should use in deciding how to vote, and what are the moral criteria a politician or a public administrator should use in deciding how to go about his own business?

Of course, social decisions also have a *game-theoretical* aspect. With no less justification, a game theorist may wish to ask the question, How can each participant, and each group of participants – each voter, or group of voters, each politician, each public servant, etc. – maximize his (or their) self-interest against other rational participants equally determined to maximize their own self-interest?

The fundamental question, however, in a normative study of social decisions is neither a moral question nor a game-theoretical one. Rather, it is a question of *organizational design*, a problem of an *optimal constitution*: It is how to design social decision-making mechanisms so as best to achieve certain social objectives or best to satisfy certain value criteria.

I think it is important to understand that this last problem, the problem of designing an optimal constitution, as such is *not* simply a *moral* problem. Rather, it is essentially an *instrumental* problem. We frame constitutions for governments, for clubs, and for associations, not in order to have these constitutions admired for their ethical or aesthetic qualities, but rather in order to maximize the chances that *good social decisions* will be generated by these constitutions. (Of course, the goodness of these social decisions in turn must be judged by how well they serve the public interest, i.e., it must be judged ultimately by moral criteria.) Only confusion can result from treating the problem of optimal constitutional design as if it were simply, or even primarily, a *moral* problem.

For example, proportional representation is presumably a 'fairer' electoral system than one based on one-legislator constituencies because it gives small parties a better chance of proper representation. But, in my opinion, this is far from being a conclusive argument in favor of proportional representation if there is some evidence (as there seems to be) that proportional representation tends to give rise to less stable and less effective governments. We simply cannot make a rational choice between proportional representation and alternative electoral systems without ascertaining which electoral system is likely to produce better social decisions and better government policies.

After this introduction, let me now ask, What are the implications of

Theorem 5 for the problem of optimal constitutional design? At first, it may appear that the theorem is bound to have absolutely revolutionary implications. Students of social decision processes have been disheartened for many years by the negative implications of Arrow's Impossibility Theorem. Now, we have at last found a decision process that completely escapes Arrow's theorem!

In actual fact, even though the implications of Theorem 5 are important, they are somewhat limited in scope. What Theorem 5 primarily shows is that maximization of a utilitarian social welfare function is a very good decision rule for *personal* moral decisions. No doubt, it is an equally good decision rule for some *social* decision processes, but only for those that are structurally suited for using maximization of a mathematical function as a decision criterion – and this means essentially social decisions delegated to *one* administrative official (or perhaps to a small group of officials), who can very well base his (or their) decisions on maximizing a social welfare function.

On the other hand, classical political theory tells us very clearly what the dangers are in delegating too much social decision-making power to individual officials (or to small groups of them). To be sure, *if* these officials are persons of great ability, and *if* the decision-making tasks delegated to them do not give rise to unmanageable problems of collecting and processing information, *then* this arrangement may produce highly efficient and altogether very satisfactory administrative decisions. Theorem 5, of course, only confirms this conclusion. But it is notoriously hard to ensure that the officials entrusted with important decisions will always be persons of superior ability. Moreover, centralized decision-making does quite often pose very grave problems in collecting and processing information. But, most important of all, any system based on great concentration of power is sooner or later bound to give rise to intolerable political abuses.

Thus, we reach the following conclusion: We can avoid Arrow's Impossibility Theorem by delegating many social decisions to public officials and by requiring them to make their decisions by maximizing a utilitarian social welfare function. But this approach has its own practical difficulties and is open to dangerous political abuses. For reasons well known to any economist, most economic decisions will be handled much better by the market mechanism than they could ever be handled by the most enlightened utilitarian public officials; whereas the basic political decisions of the society are much better left for that notoriously inefficient, quarrel-prone, and Impossibility-Theorem-ridden social-decision mechanism that we call constitutional representative democracy.

8. RULE UTILITARIANISM VS. ACT UTILITARIANISM

By Theorem 4, a rational morality must be based on maximizing the average utility level of all individuals in society, or, as I will briefly say, it must be based on *maximizing social utility*. It turns out, however, that this principle in itself does not yet uniquely determine the actual content of a rational moral code, because a mathematical maximization problem is not yet fully determined by specifying its *maximand* – unless we also specify the *constraints* of maximization (or the absence of any such constraints).

Depending on how these constraints are defined, utilitarian theory has historically appeared in two different versions, called act utilitarianism and rule utilitarianism. *Act utilitarianism* is the view that the utilitarian maximization criterion should be directly applied to each individual act: thus, this theory defines a morally right act as that particular act that, under the actual conditions, will maximize social utility. In contrast, *rule utilitariansim* is the view that the utilitarian criterion should be applied in the first instance to the moral rules, and then these moral rules should be used to direct individual acts: thus, we must first define the correct moral rule for any given situation as that particular behavioral rule that would maximize social utility if everybody followed it in similar situations; and then we can define a morally right act as an act conformable to the correct moral rule which applies to this particular type of situation.[8]

Mathematically, these two versions of utilitarian theory can be modelled as follows: Society consists of $(n + m)$ individuals. Of these latter, individuals $1, \ldots, n$ are utilitarian moral agents whose behavior is determined by utilitarian considerations. On the other hand, individuals $n + 1, \ldots, n + m$ are nonutilitarian agents, whose behavior is determined by self-interest or by some form of nonutilitarian morality (e.g., by conformity to some traditional moral code). The purpose of our model is to determine the moral strategies of the utilitarian agents. The moral strategies of the nonutilitarian agents are regarded as *given*, i.e., as being determined by forces outside our model.

Act utilitarian theory requires each utilitarian agent i ($i = 1, \ldots, n$) to choose his strategy s_i in such a way as to maximize social utility, under the assumption that all other utilitarian agents' strategies (as well as the nonutilitarian agents' strategies) are *given*. Thus, the mathematical problem we have to solve is:

Maximize $W_i(s_1, \ldots, s_i, \ldots, s_n; s_{n+1}, \ldots, s_{n+m})$, subject to the constraints

(8.1) $s_i \in S$.

(8.2) $s_j = r_j = \text{const. for } j = 1, \ldots, i - 1, i + 1, \ldots, n$.

(8.3) $s_k = t_k = \text{const. for } k = n + 1, \ldots, n + m$.

Here S is the set of all strategies available to each agent (assumed to be the same set for all agents).

In contrast, rule utilitarian theory requires each utilitarian agent i to choose a strategy s_i in such a way as to maximize social utility, under the assumption that all other utilitarian agents will follow the same set of moral rules, i.e., will follow the *same* strategy (while the nonutilitarian agents' strategies are given). Thus, the mathematical problem is:

Maximize $W_i(s_1, \ldots, s_i, \ldots, s_n; s_{n+1}, \ldots, s_{n+m})$, subject to the constraints

(8.4) $s_i \in S$.

(8.5) $s_1 = \cdots = s_i = \cdots = s_n$.

(8.6) $s_k = t_k = \text{const. for } k = n + 1, \ldots, n + m$.

It is easy to see that the two utilitarian theories will often lead to quite different moral decisions. This can be illustrated by three voting examples. In each example, the voters are asked to vote about a socially very important measure M. But voting involves some minor inconvenience. I will assume that all voters are utilitarians.

Example 1. There are 1000 voters, all in favor of measure M. But the measure will pass only if all of them actually vote. In this case, an act utilitarian voter will vote only if he is reasonably sure that *all* the other 999 voters will also vote.

If the voters follow act utilitarian theory, then this voting situation will have the nature of a *game* with 1000 players, in which every player i tries to maximize his own estimate of social utility, the social welfare function W_i. (An interesting special case is that where the players' definitions of social utility are identical, i.e., where $W_1 = \cdots = W_{1000}$. In this case we obtain a game of a very special form in which the players have identical payoff functions.) This game has *two* Nash equilibrium points: one corresponds to everybody's voting while the other corresponds to nobody's voting.

On the other hand, rule utilitarian theory permits only *one* possible outcome, where everybody votes.

Examples 2 and 3 are like Example 1, except that 'yes' votes by 800 individuals out of the 1000 voters are sufficient to pass measure M. In Example 2, the voters will be permitted only *pure* strategies while in Example 3 both *pure* and *mixed* strategies will be permitted.

In *Example 2* (as well as in Example 3), an act utilitarian voter will vote only if he is reasonably sure that *exactly* 799 other voters will vote – which means that he is very unlikely to vote. If the voters follow act utilitarian theory, then the resulting game will have a very large number of Nash equilibrium points – a number equal to the number of ways we can select 800 elements from a set of 1000 elements. (But, for reasons already stated, it is very unlikely that the outcome will be any of these equilibrium points.)

On the other hand, if the voters follow rule utilitarian theory, then *all* 1000 voters will actually vote, even though this will be a socially wasteful policy since only 800 votes are needed. But, in view of (8.5), rule utilitarian theory permits a choice only between everybody's voting and nobody's voting; and, of these two alternatives, the first will yield a higher social utility. Yet, even if a vote by all 1000 voters is wasteful, it will be preferable result over the likely outcome of act utilitarianism (which is that too few voters will vote to ensure passage of the measure).

In *Example 3*, if the voters follow act utilitarian theory, then the resulting game will have an even larger number of equilibrium points than the game obtained in Example 2 had: the pure-strategy equilibrium points obtained in Example 2 will remain equilibrium points, but they will now be joined by a very large number of mixed-strategy equilibrium points.

On the other hand, if the voters follow rule utilitarian theory, then they will all use the same mixed strategy. It can be shown that this mixed strategy will involve voting with a probability slightly larger than $800/1000 = 0.80$, and not voting with the remaining probability. (It can also be shown that this particular outcome is identical with *one* of the very large number of mixed-strategy equilibrium points of the act utilitarian game. For proof and further discussion, see Harsanyi, 1977b.)

These examples show that the rule utilitarian decision criterion is often much more effective than the act utilitarian criterion is in organizing socially desirable cooperation and strategy coordination among people of good will. This fact may be called the *coordination effect.*

It can also be shown, by using different examples, that rule utilitarianism has much more desirable effects also on people's expectations and incentives. These facts may be called the *expectation effect* and the *incentive effect* (see Harsanyi, 1977b).

Let me add that rule utilitarian theory is important not only as an ethical theory but also as an analytical tool in explaining certain facts about actual human behavior, which are hard to explain by rational-choice models that make no use of rule utilitarian theory. In particular, it can help us to explain the *voter's paradox,* i.e., to explain the fact that many intelligent people vote in large elections where the probability that their individual vote will have any effect at all is virtually nil. Rule utilitarian theory tells us that, when we wonder whether to vote or not, the right question to ask is not, How much difference would it make if *I* personally did not vote? Rather, the right question is, How much difference would it make if I and *all people in a similar situation* did not vote? Clearly, in many cases it will make no difference whatever if I for one do not vote. But if *all* people like me fail to vote – or even if *many* people like me fail to vote – this may easily decide the outcome. Thus, voting even in very large elections may not be an irrational act at all.

9 . C O N C L U S I O N

I have tried to show that the Bayesian concept of rationality is the only concept consistent with our basic intuitive criteria for rationality, and this Bayesian concept of rationality logically entails a utilitarian theory of morality. I have argued that the main objections to utilitarian theory – opposition to cardinal utility and rejection of interpersonal utility comparisons – are now completely outdated. I have also tried to show that the social welfare functions defined by utilitarian theory are not subject to Arrow's Impossibility Theorem – yet that this conclusion is more important for ethics that it is for the theory of social decision processes (though it is important for the theory of administrative decisions). Finally, I have discussed the distinction between acᴛ utilitarianism and rule utilitarianism, and have argued that the latter is a much more satisfactory ethical theory than the former is. I have also tried to show that rule utilitarian theory provides the conceptual tools needed for resolving the voter's paradox problem, which has important implications for *positive* (explanatory) theories of political behavior.

University of California, Berkeley

NOTES

[1] I want to thank the National Science Foundation for supporting this research through Grant SOC77-06394 to the Center for Research in Management Science, University of California, Berkeley.

[2] This equiprobability model for moral value judgments I first proposed in Harsanyi (1953). Qualitatively it is very similar to Rawls's 'original position' based on a 'veil of ignorance', first proposed by him in 1957. But there is an important difference. Whereas I analyse my model in terms of the Bayesian expected-utility maximization principle, Rawls analyses his in terms of the highly irrational *maximin* principle, which inevitably leads him to a rather unsatisfactory ethical theory (see Harsanyi, 1975a).

[3] To be sure, my Axiom *b* and, in particular, my assumption that people's moral preferences ought to satisfy the sure-thing principle, has been criticized by Peter Diamond (1967). But, as I have tried to show (Harsanyi, 1975b, pp. 315–318), his criticism lacks validity.

[4] But if individual *i*'s behavior satisfies appropriate *independence axioms*, then his behavior under certainty *may* be sufficient to define a cardinal utility function for him. Moreover, he may try to define a cardinal utility function for himself by *introspection*. Yet, such a purely introspective cardinal utility function, even if it could be consistently defined, would have a somewhat unclear economic meaning.

[5] It can be done if he chooses the coefficients α_{ij} without making any use of information about the 'environment' (in the sense of Arrow, 1963, p. 15), but rather by relying only on information about the various individuals' preferences, the intensities of these preferences, and other psychological parameters underlying these individuals' choice behavior.

For example, some authors choose the coefficients α_{ij} after normalizing each utility function U_i in terms of the maximal and the minimal elements of the *feasible set*, say, by setting $U_i = 1$ for the feasible alternative most preferred by individual *i*, and by setting $U_i = 0$ for the feasible alternative least preferred by him. A social welfare function based on such a normalization procedure will obviously violate Condition 3 (Independence of irrelevant alternatives) because it makes essential use of information about the feasible set (which is part of the 'environment').

In contrast, other normalization procedures will be consistent with Condition 3. For example, let us assume that each utility function U_i has asymptotically an upper bound U_i^* and a lower bound U_i^0, where U_i^* and U_i^0 are *independent* of the feasible set. Then, a normalization by setting $U_i^* = 1$ and $U_i^0 = 0$ will require no environmental information, and may possibly yield a social welfare function satisfying Condition 3. The same is true, of course, for a social welfare function making no use of any normalization procedure, but rather based on direct interpersonal utility comparisons.

[6] In addition, we may also permit linear transformations that *add* a positive or negative constant to any individual utility function U_i. The result of any such transformation will be that a positive or negative constant is added to the right-hand side of Equation (2.1) or (3.1), which can always be compensated by adding the same constant to the social welfare function W_j on the left-hand side of the equation.

[7] Some of Kalai and Schmeidler's questionable conclusions arise from the fact that they use a rather unsatisfactory version of interpersonal utility comparisons, which makes essential use of information about the 'environment'. Cf. Note 5 above.

[8] Rule utilitarianism as an ethical theory was first proposed by the economist Roy Harrod (1936). But the terms 'rule utilitarianism' and 'act utilitarianism' were coined by R. B. Brandt, a well-known moral philosopher, only in 1959 (Brandt, 1959, pp. 380 and 396).

BIBLIOGRAPHY

Anscombe, F. J. and Aumann, R. J.: 1963, 'A definition of subjective probability', Annals of Mathematical Statistics 34, pp. 199–205.

Arrow, Kenneth J.: 1963, Social Choice and Individual Values (Wiley, New York).

Arrow, Kenneth J.: 1977, 'Extended sympathy and the possibility of social choice', American Economic Review, Papers & Proc. 67, pp. 219–225.

Brandt, R. B.: 1959, Ethical Theory (Prentice-Hall, Englewood Cliffs, N.J.).

Debreu, Gerard: 1959, Theory of Value (Wiley, New York).

Diamond, Peter: 1967, 'Cardinal welfare, individualistic ethics, and interpersonal comparisons of utility', Journal of Political Economy 75, pp. 765–766.

Harrod, Roy F. (now Sir Roy): 1936, 'Utilitarianism revised', Mind 45, pp. 137–156.

Harsanyi, John C.: 1953, 'Cardinal utility in welfare economics and in the theory of risk-taking', Journal of Political Economy 61, pp. 434–435. Reprinted in Harsanyi, 1976, pp. 3–5.

Harsanyi, John C.: 1955, 'Cardinal welfare, individualistic ethics, and interpersonal comparisons of utility', Journal of Political Economy 63, pp. 309–321. Reprinted in Harsanyi, 1976, pp. 6–23.

Harsanyi, John C.: 1975a, 'Can the maximin principle serve as a basis for morality? A critique of John Rawls's theory', American Political Science Review 69, pp. 594–606. Reprinted in Harsanyi, 1976, pp. 37–63.

Harsanyi, John C.: 1975b, 'Nonlinear social welfare functions: Do welfare economists have a special exemption from Bayesian rationality?', Theory and Decision 6, pp. 311–332. Reprinted in Harsanyi, 1976, pp. 64–85.

Harsanyi, John C.: 1976, Essays on Ethics, Social Behavior, and Scientific Explanation (Reidel, Dordrecht, Holland).

Harsanyi, John C.: 1977a, Rational Behavior and Bargaining Equilibrium in Games and Social Situations (Cambridge University Press, New York).

Harsanyi, John C.: 1977b, 'Rule utilitarianism and decision theory', Erkenntnis 11, pp. 25–53.

Harsanyi, John C.: 1977c, 'Morality and the theory of rational behavior', Social Research 44, pp. 623–656.

Harsanyi, John C: 1978, 'Bayesian decision theory and utilitarian ethics', American Economic Review, Papers & Proc. 68, pp. 223–228.

Hicks, John R. (now Sir John): 1939, Value and Capital (Clarendon Press, Oxford).

Kalai, Ehud and Schmeidler, David: 1977, 'Aggregation procedure for cardinal preferences: A formulation and proof of Samuelson's impossibility conjecture', Econometrica 46, pp. 1431–38.

Samuelson, Paul A.: 1974, 'Complementarity: An essay on the 40th anniversary of the Hicks–Allen revolution in demand theory', Journal of Economic Literature 12, pp. 1255–1289.

Sen, Amartya K.: 1970, Collective Choice and Social Welfare (Holden-Day, San Francisco).

Theil, Henri: 1964, Optimal Decision Rules for Government and Industry (Rand McNally, Chicago).

Von Neumann, John and Morgenstern, Oskar: 1947, Theory of Games and Economic Behavior (Princeton University Press, Princeton, N.J.).

ERIC MASKIN

DECISION-MAKING UNDER IGNORANCE WITH IMPLICATIONS FOR SOCIAL CHOICE

ABSTRACT. A new investigation is launched into the problem of decision-making in the face of 'complete ignorance', and linked to the problem of social choice. In the first section the author introduces a set of properties which might characterize a criterion for decision-making under complete ignorance. Two of these properties are novel: 'independence of non-discriminating states', and 'weak pessimism'. The second section provides a new characterization of the so-called principle of insufficient reason. In the third part, lexicographic maximin and maximax criteria are characterized. Finally, the author's results are linked to the problem of social choice.

Several authors, [2], [3], [7], and [9] have dealt with the problem of an individual who must choose from a set of alternatives when he cannot associate a probability distribution with the possible outcomes of each alternative. The problem has been called that of decision-making under complete ignorance; presumably 'complete ignorance' captures the notion that the axioms of subjective probability cannot be fulfilled. This paper is a further investigation in that tradition. In the first section we suggest a set of properties which might characterize a criterion for decision-making under ignorance. Most of these properties are familiar, but, in particular, 'independence of non-discriminating states' and 'weak pessimism' are new in this context. The second section recapitulates some of the important decision criteria in the literature and suggests a new characterization of the so-called principle of insufficient reason. In the third part, we drop the assumption invariably made by previous authors that preferences for consequences satisfy the von Neumann-Morgenstern axioms, and characterize the so-called lexicographic maximin and maximax criteria. We also provide a new axiomatization of the ordinary maximin principle. Finally, we show that several of our results translate quite easily into the theory of social choice.

THE PROPERTIES

Let C be a consequence or 'outcome' space. C contains a subset C^* of 'sure' or 'certain' outcome as well as all finite lotteries[1] with outcomes in C^*. We

Theory and Decision 11 (1979) 319–337.   0040–5833/79/0113–0319$01.90.
Copyright © 1979 by D. Reidel Publishing Co., Dordrecht, Holland, and Boston, U.S.A.

assume that the decision-maker has a preference ordering $\succeq$ on C which satisfies the von Neumann-Morganstern axioms for decision-making under uncertainty. Let $\mathscr{U}^*$ be the family of von Neumann-Morgenstern utility representations of $\succeq$. Define a *decision* d as a map $d : \Omega \to C$ where Ω is an exhaustive list of possible states of nature. Obviously there are many ways in which nature can be described, and therefore many conceivable Ω's could apply to the same world. Following Arrow-Hurwicz [2] we shall define a *decision problem P* as a set of decisions which share a common domain $\Omega(P)$. The decision-maker solves a non-empty decision problem P by choosing a non-empty subset $\hat{P} \subseteq P$. $\hat{P}$ is interpreted as the 'choice' or 'optimal' subset of P. Let $\mathscr{P}$ be the class of all non-empty decision problems P such that P and $\Omega(P)$ are finite.[2] A *decision criterion f* is a mapping $f : \mathscr{P} \to \mathscr{P}$ such that $\forall P \in \mathscr{P}$, $f(P) \subseteq P$, $f(P) \neq \emptyset$ and such that $d(w) \sim d'(w)$[3], for all $w \in \Omega$, implies that $d \in f(P)$ if and only if $d' \in f(P)$. The following are conditions that have, at various times, been deemed reasonable properties for a decision criterion f to satisfy.

PROPERTY (1). $\forall P_1, P_2 \in \mathscr{P}, d \in P_1 \subseteq P_2 \Rightarrow [d \in f(P_2) \Rightarrow d \in f(P_1)]$

Property (1) is Sen's Property α of rationality [10].

PROPERTY (2). $\forall P_1, P_2 \in \mathscr{P}$ $[d, d^1 \in f(P_1)$ and $P_1 \subseteq P_2]$ $\Rightarrow$ $[d \in f(P_2) \Leftrightarrow d^1 \in f(P_2)]$.

Property (2) is Sen's Property β and Milnor's 'row adjunction'.[9].

Together (1) and (2) constitute the Arrow-Hurwicz Property A, and, as Herzberger [6] has shown, imply that, for every Ω, f induces an ordering $\succeq_\Omega^*$ on $D_\Omega = \{d \mid \text{domain of } d = \Omega\}$ such that for any P with $\Omega(P) = \Omega$, $d^* \in f(P) \Leftrightarrow d^* \in P$ and $d^* \succeq_\Omega^* d$ for all $d \in P$.

PROPERTY (3). $\forall P_1, P_2 \in \mathscr{P}, d \in P_1 \subseteq P_2 \Rightarrow [d \in f(P_1), d \notin f(P_2) \Rightarrow f(P_2) \backslash P_1 \neq \emptyset]$.

Properties (1) and (3) together are quivalent to Luce's and Raiffa's Axiom $7'$ and Chernoff's [3] Postulate 4. (1) and (3) combined are somewhat weaker than the combination of (1) and (2).

PROPERTY (4). $\forall P \in \mathscr{P}$ $d, d^1 \in P$, if $d \in f(P)$ and $d^1(w) \succeq d(w)$ for all $w \in \Omega(P)$, then $d^1 \in f(P)$.

Property (4) is the weakest form of the domination principle. It is Arrow-Hurwicz Property D.

PROPERTY (5). $\forall P \in \mathscr{P}$ $\forall d^1 \in P$ $\forall d \in f(P)$, if $d(w) \succ d^1(w)$ for all $w \in \Omega(P)$, then $d^1 \notin f(P)$.

Property (5) is another rather weak version of domination. It is Milnor's 'Strong Domination' property.

PROPERTY (6). $\forall P \in \mathscr{P}$ $\forall d$, $d^1 \in P$ if $\forall w \in \Omega(P)$, $d(w) \succsim d^1(w)$ and $\exists w_0 \in \Omega(P)$ such that $d(w_0) \succ d^1(w_0)$, then $d^1 \notin f(P)$.

Property (6) is the usual admissibility condition. It is obviously stronger than property (5). Combined with continuity (see below), it is also stronger than property (4).

PROPERTY (7). $\forall P_1$, $P_2 \in \mathscr{P}$ such that $\Omega(P_1) = \Omega(P_2) = \Omega$, if, for some $u \in \mathscr{U}^*$, there exist $k \in \mathscr{P}$, $w_0 \in \Omega$, and bijection $h : P_1 \to P_2$ such that

$$\forall d \in P_1, u(h(d)(w)) = \begin{cases} u(d(w)) + k, & w = w_0, \\ u(d(w)), & w \neq w_0. \end{cases}$$

then, $d \in f(P_1)$ if and only if $h(d) \in f(P_2)$.

Property (7) is Milnor's column linearity condition. It amounts to demanding that if two decision problems are isomorphic except that in one, the utility derived from any decision if a certain state of nature w_0 prevails is uniformly higher than the utility from the corresponding decision in the other problem when w_0 arises, then if a given decision is optimal in the other.

PROPERTY (8). $\forall P_1$, $P_2 \in \mathscr{P}$ such that $\Omega(P_1) = \Omega(P_2)$ and $|P_1| = |P_2|$, if for some $u \in \mathscr{U}^*$, there exists a bijection $g : P_1 \to P_2$ such that for some $a > 0$, $b \in \mathscr{P}$, $u(g(d)(w)) = au(d(w)) + b$ for all $d \in P_1$ and $w \in \Omega$, then $d \in f(P_1) \Leftrightarrow g(d) \in f(P_2)$.

Property (8) is Milnor's linearity condition.

PROPERTY (9). $\forall P \mathscr{P}$ $\forall d_1, d_2, d \in P$, if $\exists u \in \mathscr{U}^*$ such that $u \circ d = \frac{1}{2} u \circ d_1 + \frac{1}{2} u \circ d_2$, then $d_1, d_2 \in f(P) \Rightarrow d \in f(P)$.

Property (9) is Milnor's convexity condition.

PROPERTY (10). Consider a sequence $\{P_i\} \subseteq \mathscr{P}$ and $P \in \mathscr{P}$. Suppose that for all i, $\Omega(P_i) = \Omega(P)$ and $|P_i| = |P| = n$. Write $P = \{d_1, \cdots, d_n\}$, $P^i = \{d_1^i, \cdots, d_n^i\}$. Then, if $\exists u \in \mathscr{U}^*$ such that $\forall j \forall w \in \Omega(P) \lim_{i \to \infty} u(d_j^i(w)) = u(d_j(w))$,

$d_j^i(P^i) \in f(P^i)$ for all i implies $d_j\, f(P)$.

Property (10) is Milnors's continuity axion.

Up to this point, all of the stated properties are arguably reasonable, but none embodies the idea of ignorance. Properties 11–13 are an attempt to capture this notion.

PROPTERTY (11). Suppose there exists a bijection $h:\Omega^1 \to \Omega$. For P with $\Omega(P) = \Omega$, define P^1 with $\Omega(P^1) = \Omega^1$ as

$$P^1 = \{d^1 \mid d^1 = d \circ h \text{ for } d \in P\}.$$

Then, $d \in f(P)$ if and only if $d \circ h \in f(P^1)$.

Property (11) is the Arrow-Hurwicz Property B and the Milnor Symmetry condition. It insists that the labelling of states and decisions be irrelevant for the decision criterion.

Consider $P_1, P_2 \in \mathscr{P}$. Following Arrow-Hurwicz, P_2 is said to be derived from P_1 by deletion of repetitious states $(P_1 \to P_2)$ if $\Omega(P_2) \subseteq \Omega(P_1)$ and if there exists a bijection $h:P_1 \to P_2$ such that $\forall w \in \Omega(P_2)\ h(d)(w) = d(w)$ and such that $\forall w \in \Omega(P_1)/\Omega(P_2), \exists w^1 \in \Omega(P_2)$ with $d(w) = d(w^1)$ for all $d \in P_1$.

PROPERTY (12). $\forall P_1, P_2 \in \mathscr{P}$, if $P_1 \to P_2$ via bijection h, then $h(d) \in f(P_2) \Leftrightarrow d \in f(P_1)$.

Property (12) is the Arrow-Hurwicz Property C and the Milnor 'Deletion of Repetitious States'. More than any other property, it captures the idea of complete ignorance, for, in effect, it asserts that dividing a state into several substates should have no effect on the chosen decision. The next condition is just a weakened version of Property (12).

PROPERTY (13). $\forall P_1, P_2 \in \mathscr{P}$, if $P_1 \to P_2$ via bijection h and if for all $d_1, d_2 \in P_1$ with $d_1 \neq d_2$, $\forall w, w^1 \in \Omega(P_1)$, not $d_1(w) \sim d_2(w^1)$, then $h(d) \in f(P_2) \Leftrightarrow d \in f((P_1)$.

PROPERTY (14). Consider $P_1, P_2 \in \mathscr{P}$ with $\Omega(P_2) \supseteq \Omega(P_1)$ and a surjection $g:P_1 \to P_2$ such that $\forall d \in P_1\ \forall w \in \Omega(P_1)\ g(d)(w) = d(w)$. Then, if for $d, d^1 \in f(P_1)$, $g(d)(w) \sim g(d^1)(w)$ for all $w \in \Omega(P_2)/\Omega(P_1)$, $g(d) \in f(P_2) \Leftrightarrow g(d^1) \in f(P_2)$.

This last property requires that adding additional states for which all decisions are equivalent does not affect the choice of optimal decisions. In effect this requirement is the strong separability axiom of Debreu [5]. Although it has not previously appeared in discussions of decision-making under ignorance, entirely analogous properties have been used recently in the social choice literature under the names of 'elimination of indifferent individuals' [4] and 'unanimity' [10].

II. THE DECISION CRITERIA

We may now state the results for the case where preferences obey the von Neumann-Morgenstern axioms.

THEOREM 1. (Arrow-Hurwicz): A decision criterion f satisfies properties (1), (2), (4), (11), (12) if and only if for each $u \in \mathscr{U}^*$ there exists a weak ordering $\succsim_u^*$ in the space of real ordered pairs (M, m) with $m \leqslant M$ such that

(a) $M_1 \geqslant M_2$ and $m_1 \geqslant m_2$ implies that $(M_1, m_1) \succsim_u (M_2, m_2)$,

(b) $\forall P \; \mathscr{P}$,

$$f(P) = \{d \in P | (\max u(d(w)), \min u(d(w)) \succsim_u^* (\max u(d'(w)),$$
$$\min u(d'(w))) \quad \text{for all} \quad d' \in P\}.$$

DEFINITION. A criterion f is the Hurwicz α-criterion for $\alpha \in [0, 1]$ if $\forall P \in \mathscr{P} \; \forall u \in \mathscr{U}$, $d^* \in f(P)$ if and only if $\alpha \max_w u(d^*(w)) + (1 - \alpha) \min_w u(d^*(w)) \geqslant \max_w u(d(w)) + (1 - \alpha) \min_w u(d(w))$ for all $d \in P$.

THEOREM 2. A decision criterion f satisfies properties (1), (2), (4), (5), (8), (11), (12) if and only if $\exists \alpha \in [0, 1]$ such that $\forall P \in \mathscr{P}, f(P) \subseteq f^\alpha(P)$ where f^α is the Hurwicz α-criterion.

Remark. It should be noted that this theorem does not require continuity (property (10)). If, however, continuity is also stipulated, we obtain Theorem 3 (see below).

Proof. If f satisfies the stipulated properties, we may apply Theorem 1 and, for choice of $u \in \mathscr{U}^*$, define an ordering $\succsim_u^*$ as above. Following Milnor's argument, let α_u be the supremum of all $\alpha^1 \in \mathbb{R}$ such that $(1, 0) \succsim_u^*(\alpha', \alpha')$. By property (5), $0 \leqslant \alpha_u \leqslant 1$. Clearly $(1,0) \succ_u^*(\alpha', \alpha')$ if $\alpha' < \alpha_u$, and

$(\alpha', \alpha') \succ_u^* (1, 0)$ if $\alpha' > \alpha_u$. By property (8) and the fact that all representations of $\succsim$ differ by positive linear transformations, α_u does not depend
on u, and we may consequently delete its subscript. By property (8),
$(M, m) \succsim_u^* (\alpha'M + (1 - \alpha')m, \alpha'M + (1 - \alpha')m)$ if $0 \leqslant \alpha' < \alpha$, and $(\alpha'M +
(1 - \alpha')m, \alpha'M + (1 - \alpha')m) \succ_u^* (M, m)$ if $\alpha < \alpha' \leqslant 1$. Suppose that for some
$P \in \mathscr{P} \exists d^* \in f(P)$ and $\exists d \in P$ such that $\alpha M^* + (1 - \alpha)m^* < \alpha M^0 + (1 - \alpha)m^0$,
where $M^* = \max_w (u(d^*(w))$, $m^* = \min_w u(d^*(w))$, $M^0 = \max_w u(d(w))$,
$m^0 = \min_w u(d(w))$. Choose sequence of real numbers $\{\epsilon_i\}$ and $\{\delta_i\}$ such that
(a) $\lim_{i \to \infty} \epsilon_i = 0$, (b) $\forall i\ \epsilon_i > 0$ if $\alpha < 1$ and $\epsilon_i = 0$ if $\alpha = 1$, (c) $\lim_{i \to \infty} \delta_i = 0$, and
(d) $\forall i\ \delta_i > 0$ if $\alpha > 0$ and $\delta_i = 0$ if $\alpha = 0$.

For sufficiently large i, $((\alpha + \epsilon_i)M^* + (1 - \alpha - \epsilon_i)m^*, (\alpha + \epsilon_i)M^* + (1 -
\alpha - \epsilon)m^*) \succsim_u^* (M^*, m^*) \succ_u^* (M^0, m^0) \succsim_u^* ((\alpha - \delta_i)M^0 + (1 - \alpha + \delta_i)m^0,
(\alpha - \delta_i)M^0 + (1 - \alpha + \delta_i)m^0)$. By definition of $\succsim_u^*$, $(\alpha + \epsilon_i)M^* + (1 - \alpha -
\epsilon_i)m^* (\alpha - \delta_i)M^0 + (1 - \alpha + \delta_i)m^0$. Therefore, $\alpha M^* + (1 - \alpha)m^* \geqslant \alpha M^0 +
(1 - \alpha)m^0$, a contradiction. The other direction of implication is trivial.

$$Q.E.D.$$

THEOREM 3 (Milnor). A criterion f satisfies properties (1), (2), (5), (8),
(10), (11), (12) if and only if $\exists \alpha \in [0, 1]$ such that f is the Hurwicz α-criterion.

DEFINITION. Criterion f is the maximin criterion if and only if, $\forall P \in \mathscr{P}$
$\forall u \in \mathscr{U}^*$, $d^* \in f(P)$ if and only if $\min_w u(d^*(w)) \geqslant \min_w u(d(w))$ for all $d \in P$.

THEOREM 4 (Milnor). A criterion f satisfies properties (1), (2), (5), (9),
(10), (11), (12) if and only if f is the maximin criterion.

THEOREM 5. A criterion f satisfies properties (1), (2), (4), (5), (9), (11),
(12) if and only if $\forall P \in \mathscr{P}$, $f(P) \subseteq f^*(P)$ where f^* is the maximin criterion.
    Proof. The proof is identical to Milnor's proof of Theorem 4, except for
a minor alteration forced by lack of continuity.

DEFINITION. f is the principle of insufficient reason (the Laplace criterion)
if $\forall P \in \mathscr{P}\ \forall u \in \mathscr{U}^*$, $d^* \in f(P)$ if and only if $\Sigma_{w \in \Omega(P)}\ u(d^*(w)) \geqslant
\Sigma_{w \in \Omega P}\ u(d(w))$ for all $d \in P$.

THEOREM 6 (Chernoff). Criterion f satisfies properties (1), (3), (4), (6), (7), (9), (11) if and only if f is the principle of insufficient reason.

THEOREM 7 (Milnor). Criterion f satisfies properties (1), (2), (5), (7), (11) if and only if f is the principle of insufficient reason.

THEOREM 8. A criterion f satisfies properties (1), (2), (6), (10), (11), (14) if and only if f is the principle of insufficient reason.

Proof. Choose $\Omega = \{w_1, \ldots, w_n\}$ and $u \in \mathscr{U}^*$. Let $D_\Omega = \{d \mid$ domain of $d = \Omega\}$. (1) and (2) imply that there exists an ordering $\succsim_\Omega^*$ on D_Ω such that $\forall P \in \mathscr{P}$ with $\Omega(P) = \Omega$, $d^* \in f(P) \leftrightarrow d^* \succsim_\Omega^* d$ for all $d \in P$. By definition of a decision criterion, if $u(d(w)) = u(d'(w))$ for all $w \in \Omega$ and $d, d' \in D_\Omega$ then $d \sim_\Omega^* d'$. Therefore, if we write $\Omega = (w_1, \ldots, w_n)$, $\succsim_\Omega^*$ induces an ordering $\succsim_\Omega$ of $(u(C))^n$ such that $x \succsim_\Omega y$ if and only if $d_x \succsim_\Omega^* d_y$ for d_x, $d_y \in D_\Omega$ such that $x = (u(d_x(w_1)), \ldots, u(d_x(w_n))$, $y = (u(d_y(w_1)), \ldots, u(d_y(w_n)))$. Since C contains all finite lotteries of outcomes in C^*, $u(C)$ is connected. Consider $A_{x_0} = \{x \in (u(C))^n | x_0 \succsim_\Omega x\}$ for some $x_0 \in (u(C))^n$. Choose a convergent sequence $\{x_i\} \subseteq A_{x_0}$ such that $\forall_i, x_0 \succsim_\Omega x_i$. Take $x_\infty = \lim_{i \to \infty} x_i$. Choose a sequence $\{P_i\} \subseteq \mathscr{P}$ and $P_\infty \in \mathscr{P}$ with $\Omega(P_i) = \Omega(P_\infty) = \Omega$ for all i, such that $P_i = \{d_i, d_0\}$ $P_\infty = \{d_\infty, d_0\}$, $(u(d_i(w_1)), \ldots, u(d_i(w_n))) = x_i$, $(u(d_0(w_1)), \ldots, u(d_0(w_n))) = x_0$, $(u(d_\infty(w_1)), \ldots, u(d_\infty(w_n))) = x_\infty$. Since $x_0 \succsim_\Omega x_i$, $d_0 \in f(P_i)$ for all i. By (10), $d_0 \in f(P_\infty)$. Therefore $x_0 \succsim_\Omega x_\infty$. So, A_{x_0} is closed. Similarly $B_{x_0} = \{x \in (u(C))^n | x \succsim_\Omega^* x_0\}$ is closed for any $x_0 \in (u(C))^n$. By (14), one may easily show that the ordering induced by $\succsim_\Omega$ on $\mathbb{R}^{n-m}$ by fixing m components of the vectors in $(u(C))^n$ is independent of the values at which they are fixed. Therefore, the hypotheses of Debreu's Theorem [5] are satisfied, and we conclude that there exist continuous functions $g_1, g_2, \ldots, g_n \colon u(C) \to \mathbb{R}$ such that $\forall d, d^1 \in D_\Omega$

$$(1) \qquad d \sim_\Omega^* d^1 \quad \text{if and only if} \quad \sum_{i=1}^{n} g_i(u(d(w_i))) \geqslant \sum_{i=1}^{n} g_i(u(d^1(w))).$$

By (11) all the g_i's are equal to some continuous $g \colon (u(C)) \to \mathbb{R}$. By (6), g is strictly increasing. By (8) we may use the argument of Maskin [8] to conclude that $g(u)$ is a positive linear transformation of u. This establishes the theorem.

III. DECISION-MAKING WITHOUT THE VON NEUMANN-MORGENSTERN AXIOMS

It is perhaps a bit odd to insist that individual preferences obey a set of probabilistic axioms in order to develop a theory which rejects the use of probabilities. In this section, we drop the assumption that $\succsim$ satisfies the von Neumann-Morgenstern axioms and that C need contain all finite lotteries of sure outcomes. For convenience we shall assume that $\succsim$ is representable by a class $\mathcal{U}$ of real-valued utility functions on C.[4] Obviously, for $u \in \mathcal{U}$, any monotone increasing transformation of u is also in $\mathcal{U}$. Properties (1)–(6), (11), (12), (13), (14) remain the same in this framework as before. Properties (7)–(10) can be modified by substituting $\mathcal{U}$ for $\mathcal{U}^*$. Properties (7), (8), and (9), of course, now make no intuitive sense. However, (9) can be modified appropriately in the following obvious way.

PROPERTY (9'). $\forall P \in \mathcal{P} \forall d_1, d_2, d \in P$, if, for each $w \in \Omega(P)$, $d_1(w) \succsim d(w) \succ d_2(w)$, or $d_2(w) \succsim d(w) \succ d_1(w)$, or $d_1(w) \sim d(w) \sim d(w)$, then $d_1, d_2 \in f(P)$ implies that $d \in f(P)$.

Among the results of the previous section, Theorems 2, 3, and 6–8 will not hold in this new context because they depend on all representations of $\succsim$ being linear transformations of one another. Theorem 1 will carry over, but Theorems 4 and 5, as indeed several other theorems to follow, depend crucially on the number and distribution of indifference classes of $\mathcal{U}$. When the von Neumann-Morgenstern axioms are assumed, this is no problem because whenever there are at least two distinct indifference classes, there is a continuum of them. Without these axioms, however, we may run into trouble. Let us, for example, examine Milnor's proof of Theorem 4. For utility function u, he first observes that because of Theorem 1, a decision $d: \Omega \to C$ may, for the purposes of ranking the decisions in D, be identified with the pair (m, M), where $m = \min u(d(w))$ and $M = \max u(d(w))$. He then considers the matrix

	w_1	w_2	w_3
$u \circ d_1$	m	$\frac{1}{2}(m + M)$	$\frac{1}{2}(m + M)$
$u \circ d_2$	m	m	M
$u \circ d_3$	m	M	m

By using this matrix, he implicitly assumes that there exists a consequence x in C such that $u(x) = \frac{1}{2}(m + M)$, an assumption validated by the von Neumann-Morgenstern axioms. In our present context, however, we must make the following supposition.

DENSENESS ASSUMPTION. For all $x, y \in C$, such that $x \succ y$, $\exists z \in C$ for which $x \succ z \succ y$.

THEOREM 9. If $\succsim$ satisfies the Denseness Assumption, then a criterion f satisfies properties (1), (2), (5), (9'), (10), (11), and (12) if and only if f is the maximin criterion.

Proof. The proof is an adaption of Milnor's proof of Theorem 4. Choose $u \in \mathcal{U}$. Suppose $\succsim$ satisfies the Denseness Assumption and f satisfies the hypothesized properties. For any Ω, let $\succsim_\Omega^*$ be the ordering induced by f on D_Ω. Consider $d \in D_\Omega$ such that $M > m$ where $M = \max u(d(w))$ and $m = \min u(d(w))$. Take $k_0 = \inf\{k \mid k \in u(C), k > m\}$. The following argument will demonstrate that we may assume that $k_0 = m$. Suppose instead that $k_0 > m$. If there exists $x_0 \in C$ such that $u(x_0) = k_0$, then, by the Denseness Assumption, $\exists y_0 \in C$ such that $k_0 > u(y_0) > m$, a contradiction of k_0. Therefore, there does not exist $x_0 \in C$ such that $u(x_0) = k_0$.

Take
$$u_0(x) = \begin{cases} u(x), u(x) \leqslant m, \\ u(x) - k_0 + m, u(x) > m. \end{cases}$$

Clearly $u_0 \in \mathcal{U}$ and $\inf\{k \mid k \in u_0(C), k > m\} = m$.

So we may replace u with u_0, thereby allowing us to take $k_0 = m$. Since $k_0 = m$, we may choose a sequence $\{k_n\} \subseteq u(C) \cap (m, M)$ such that $\lim\limits_{n \to \infty} k_n = m$. Choose $W = \{w_1, w_2, w_3\} \subseteq \Omega^5$ and decisions $\bar{d}, \bar{\bar{d}}, d_0, d_n$: $\Omega \to C$ such that

$$u(\bar{d}(w_1)) = m, u(\bar{d}(w_2)) = m, u(\bar{d}(w_3)) = M, u(\bar{d}(w)) = m$$

for $w \notin W$

$$u(\bar{\bar{d}}(w_1)) = m, u(\bar{\bar{d}}(w_2)) = M, u(\bar{\bar{d}}(w_3)) = m, u(\bar{\bar{d}}(w)) = m$$

for $w \notin W$

$$u(d_0(w)) = m \text{ for all } w \in \Omega,$$

and for all n, $u(d_n(w_1)) = m$, $u(d_n(w_2)) = k_n$, $u(d_n(w_3)) = k_n$, $u(d_n(w)) = m$ for $w \notin W$.

By Property 11, $\bar{d} \sim_\Omega^* \bar{\bar{d}}$. By convexity ($(9')$), for all n, $\bar{d} \sim_\Omega^* \bar{\bar{d}}$. Strong domination ($(5)$) and continuity ($(10)$) together imply that f satisfies property (4). Therefore, $\bar{d} \sim_\Omega^* \bar{\bar{d}} \sim_\Omega^* d$. By convexity ($(9')$), $d_n \sim_\Omega^* \bar{d}$ for all n. Hence, by continuity ((10)), $d_0 \sim_\Omega^* \bar{d} \sim_\Omega^* d_0$. f is clearly the maximum criterion.

Q.E.D.

Our next series of results does not require the Denseness Assumption. The theorems do, however, necessitate the following weaker hypothesis.

COUNTABILITY ASSUMPTION. Either $\gtrsim$ has only a single indifference class, or it has at least countably infinitely many.

We shall need to add to our list two additional properties. The first is a strengthening of property (8). The second merely states that the decision-maker does not always choose as if the best possible outcome will occur.

PROPERTY (15). For $P, P' \in \mathscr{P}$ such that $\Omega(P) = \Omega(P')$ and $|P| = |P'|$, suppose there exists a $\gtrsim$-preserving bijection[6] $\gamma : P \to P'$. Then $d \in f(P)$ if and only if $\gamma(d) \in f(P')$.

PROPERTY (16). (Weak Pessimism): There exist $P \in \mathscr{P}$, $d \in f(P)$, $d' \in P\backslash f(P)$ and $u \in \mathscr{U}$ such that $\max_{w \in \Omega(P)} u(d(w)) < \max_{w \in \Omega(P)} u(d'(w))$.
　　For any d with domain Ω and $w_0 \in \Omega$, let $D^d(w_0) =$
$\{w \in \Omega | d(w_0) \gtrsim d(w)\}$.

DEFINITION. A criterion f is the lexicographic maximin if and only if for any $P \in \mathscr{P}$, $d \in f(P)$ implies that there do not exist $d^* \in P$, $w_0 \in \Omega$ and permutation $g : \Omega \to \Omega$ such that $d^*(w_0) > d(g(w_0))$ and $d^*(w) \sim d(g(w))$ for all $w \in D^{d^*}(w_0)$.
　　The lexicographic maximax criterion is defined in the obvious analogous way.

THEOREM 10. If $\gtrsim$ satisfies the Countability Assumption, then a criterion f which satisfies properties (1), (2), (6), (11), (14), (15) must be either the lexicographic maximin or lexicographic maximax.

THEOREM 11. If $\succsim$ satisfies the Countability Assumption, then a criterion f which satisfies properties (1), (2), (6), (11), (14), (15) and (16) must be the lexicographic maximin.

Observe that neither Theorem 10 or 11 invokes property (13). The theorems are proved by demonstrating that their statements can be translated into the language of social choice and by then applying a result due to d'Aspremont and Gevers [4]. We must first develop the necessary social choice terminology. Following Maskin [8], let $N = \{1, \ldots, n\}$ be a set of individuals who constitute society and let X be a set of social alternatives. Let $\mathscr{R}$ be the set of all orderings of X and $\mathscr{V}$ the set of all bounded **R** -valued functions on $X \times N$. For $v \in \mathscr{V}$, $v(x, i)$ is the utility that the ith individual derives from alternative x. A social welfare functional (SWFL) g is a mapping $g : \mathscr{V} \to \mathscr{R}$. The following are possible properties which a SWFL g may possess.

INDEPENDENCE. $\forall v_1, v_2 \in \mathscr{V}$ $\forall B \subseteq X$, if $v_1(x, .) = v_2(x, .)$ for all $x \in B$, then $g(v_1)$ and $g(v_2)$ coincide on B.

STRONG PARETO PROPERTY. $\forall x, y \in X \ \forall v \in \mathscr{V}$, if $\forall i \in N$, $v(x, i) \geqslant v(y, i)$ and $\exists j \in N$ such that $v(x, j) > v(y, j)$, then $x \, P \, y$ where P is the strong ordering corresponding to $g(v)$.

ANONYMITY. For any permutation σ of N, if for $v_1, v_2 \in \mathscr{V} \ \forall i \in N$, $\forall x \in X \ v_1(x, i) = v_2(x, \sigma(i))$, then $g(v_1) = g(v_2)$.

ELIMINATION OF INDIFFERENT INDIVIDUALS. $\forall v_1, v_2 \in \mathscr{V}$, if $\exists M \subseteq N$ such that $\forall i \in M$, $v_1(\cdot, i) = v_2(\cdot, i)$ while $\forall j \in N \backslash M$, $\forall x, y \in X$, $v_1(x, j) = v_1(y, j)$ and $v_2(x, j) = v_2(y, j)$, then $g(v_1) = g(v_2)$.

COORDINALITY. Consider $v_1, v_2 \in \mathscr{V}$ such that $v_1 = \emptyset(v_2)$ where $\emptyset$ is a strictly monotone increasing function. Then a SWFL g satisfies coordinality if $g(v_1) = g(v_2)$.

Above, we defined $\mathscr{V}$ as containing *all* bounded **R** -valued functions on $X \times N$. Because coordinality is assumed, however, one can modify arguments due to d'Aspremont and Gevers to obtain

THEOREM 12. For $N = \{1, 2, \ldots, n\}$ and a set of social alternatives X containing at least three elements, let $\mathscr{V}$ be a set of bounded $\mathbb{R}$ -valued functions of $X \times N$ which is sufficiently large to induce all orderings of $X \times N$.[7] Then a SWFL $g: \mathscr{V} \to \mathscr{R}$ which satisfies independence, the strong Pareto property, anonymity, coordinality, and elimination of indifferent individuals is either the lexicographic maximin or maximix principle.

Proofs of Theorems 10 and 11. Suppose f satisfies the hypothesized properties. Choose $u \in \mathscr{U}$. For a given nature space $\Omega = \{w_1, \ldots, w_n\}$, let each state $w_i \in \Omega$ be interpreted as an individual i. Take $N = \{1, 2, \ldots, n\}$. Choose a positive integer m, and select a decision problem $P_0 \in \mathscr{P}$ such that $|P_0| = m$ and $\Omega(P_0) = \Omega$. Let $\mathscr{P}_\Omega^m = \{P \in \mathscr{P} \mid \Omega(P) = \Omega, |P| = m\}$. For each $P \in \mathscr{P}_\Omega^m$, select a bijection $h_P : P_0 \to P$. Write $P_0 = \{d_1, \ldots, d_m\}$. Associate with each d_j a social alternative x^j. Take $X_\Omega^m = \{x^1, \ldots, x^m\}$. A decision problem $P \in \mathscr{P}_\Omega^m$ induces an interpersonal utility function $v_P : X_\Omega^m \times N_\Omega \to \mathbb{R}$ where

$$v_P(x^j, i) = u(h_P(d_j)(w_i)) \quad \text{for all} \quad i, j$$

$v_P(x^j, i)$ is interpreted as the utility that individual i derives from alternative x^j. For given $\Omega, \mathscr{P}_\Omega^m, P_0$, and set of bijections $\{h_P \mid h_P : P_0 \to P, P \in \mathscr{P}_\Omega^m\}$, the decision criterion f induces a SWFL $g_\Omega^m : \mathscr{V}_\Omega^m \to \mathscr{R}_\Omega^m$, where $\mathscr{V}_\Omega^m = \{v_P \mid P \in \mathscr{P}_\Omega^m\}$ and $\mathscr{R}_\Omega^m$ is the set of all orderings of X^m, via the following relation

$$\forall v_P \in \mathscr{V}_\Omega^m \, \forall x^k, x^l \in X_\Omega^m, x^k g_\Omega^m(v_P) x^l \text{ iff } h_P(d_k) \succsim_\Omega^* h_P(d_l),$$

where $\succsim_\Omega^*$ is the ordering induced by f on D_Ω. ($\succsim_\Omega^*$ exists since f satisfies properties (1) and (2)). Because f satisfies properties (1) and (2), g_Ω^m is clearly well-defined and satisfies independence. Since f satisfies property (6), g_Ω^m satisfies the Pareto property. Because f satisfies property (11), g_Ω^m is obviously anonymous. From property (14) g_Ω^m satisfies the elimination of indifferent individuals property. Property (15) clearly translates into coordinality. It should be observed that the set $\mathscr{V}_\Omega^m$ may not include all bounded $\mathbb{R}$ -valued functions on $X_\Omega^m \times N_\Omega$. However, by the Countability Assumption $\succsim$ has more than nm indifference classes. Thus $\mathscr{V}_\Omega^m$ will induce all possible orderings of $X_\Omega^m \times N_\Omega$. Thus $\forall \Omega, \forall m, g_\Omega^m$ satisfies all the hypotheses of Theorem 12 and must therefore be either the lexixographic maximin or maximax. Translating back into the language of decision-making under ignorance, f must be either the lexicographic maximin or maximax. Thus Theorem 10 is established. To establish Theorem 11, we note that property (16) translates into the statement that for some $\bar{\Omega}, \bar{m},$

$$(*) \quad \begin{cases} \exists v \in \mathscr{V}_{\bar{\Omega}}^{\bar{m}} \; \exists x, y \in X_{\bar{m}} \text{ such that} \\ \max_i v(x, i) < \max_i v(y, i) \text{ but } x g_{\bar{\Omega}}^{\bar{m}}(v) y. \end{cases}$$

Now $g_{\bar{\Omega}}^{\bar{m}}$ is, by the above arguments, either the lexicographic maximin or maximax. But by (*), it cannot be the lexicographic maximax. Therefore, $g_{\bar{\Omega}}^{\bar{m}}$ is the lexicographic maximin. But since f must be either the lexicographic maximin or maximax, this in turn implies that f is actually the lexicographic maximin.

Q.E.D.

It is well known that when a domain of individual preferences is restricted, the set of social welfare functions which satisfy a given list of properties is enlarged. For example, restricting the domain of preferences often enables one to define social welfare functions satisfying all of Arrow's properties [1], although no such SWF exists for the unrestricted domain. As we have seen, positing a minimum number of indifference classes of $\succsim$ is essential to show that the corresponding SWF has unrestricted domain. Therefore, limiting the number of indifference classes is equivalent to restricting the corresponding domain of preferences. One would expect, then, that the properties of Theorems 10 and 11 might not be sufficient to uniquely characterize the lexicographic maximin if a minimum number of indifference classes is lacking. This conjecture is validated by the following example.

Suppose that $\succsim$ has only three indifference classes and that for $x, y, z \in C$, $z \succ y \succ x$. Let f^* be the criterion such that, for any $\Omega, d, d' \in D_\Omega, d \succsim {}^*_\Omega d'$, if and only if $d \succsim {}^{Lm}_\Omega d'$, where $\succsim {}^*_\Omega$ and $\succsim {}^{Lm}_\Omega$ are the orderings induced on D by, respectively, f^* and the lexicographic maximin, unless $\exists w_1, w_2, w_3 \in \Omega$ and $\exists$ some permutation σ of Ω such that $d'(w_1) \sim x, d'(w_2) \sim z, d'(w_3) \sim z$, $d(\sigma(w_1)) \sim y, d(\sigma(w_2)) \sim y, d(\sigma(w_3)) \sim y$, and $d'(w) \sim d(\sigma(w))$ for all $w \in \Omega \setminus \{w_1, w_2, w_3\}$ in which case $d' \succ {}^*_\Omega d$. To check that f^* indeed induces a transitive ordering of D_Ω, consider $\Omega = \{w_1, w_2, w_3\}$ and $d, d' \in D_\Omega$ such that

	w_1	w_2	w_3
d'	x	z	z
d	y	y	y

By hypothesis, $d' \succ^*_\Omega d$. If there are any intransitivities in $\succsim^*_\Omega$, it must be because $\exists \bar{d}, \bar{\bar{d}} \in D_\Omega$ such that $\bar{d} \succsim^*_\Omega d'$ and $d \succsim^*_\Omega \bar{\bar{d}}$ but $\bar{d} \succsim^*_\Omega \bar{\bar{d}}$. Now if $\bar{d} \succsim^*_\Omega d'$, either there exists a permutation of $\{1, 2, 3\}$ such that $\bar{d}(w_{\sigma(i)}) \sim d'(w_i)$ for all $i \in \{1, 2, 3\}$, or, for all i, $\bar{d}(w_i) \succsim y$. If $d \succsim^*_\Omega \bar{\bar{d}}$, then either $\bar{\bar{d}}(w_i) \sim y$ for all $i \in \{1, 2, 3\}$ or there exists $i \in \{1, 2, 3\}$ such that $\bar{\bar{d}}(w_i) \sim x$. It is straightforward to check that in all cases $\bar{d} \succ^*_\Omega \bar{\bar{d}}$ as desired. Thus f^* satisfies properties (1), (2), (6), (11), (14), (15), and (16), but is obviously not the lexicographic maximin. Clearly, additional properties must be hypothesized to obtain a result like Theorem 11 without stipulating the Countability Assumption. It turns out that actually only one additional property must be added to the list: property (13), the weakened form of (12). We shall assume from now on that $\succsim$ has at least 6 indifference classes. If this assumption is not met, the proofs of Theorems 13–15 are even simpler.

THEOREM 13. A criterion f which satisfies properties (1), (2), (6), (11), (13), (14), (15) is either the lexicographic maximin or lexicographic maximax.

   Proof. Consider a decision criterion f which satisfies the above axioms. Let $P^0 = \{d^0_1, d^0_2\}$ and $\Omega(P^0) = \Omega^0 = \{w_1, w_2\}$ for which $d^0_1(w_1) \succ d^0_2(w_1) \succ d^0_2(w_2) \succ d^0_1(w_2)$. There are three possible cases

(I) $d_1 \notin f(P^0), d_2 \in f(P^0)$

(II) $d_1 \in f(P^0), d_2 \notin f(P^0)$

(III) $d_1, d_2 \in f(P^0)$

By axioms (11) and (15), if (I) holds, then for any $\tilde{\Omega} = \{w_1, w_2\}$, $\tilde{P} = \{d_1, d_2\}$, $\Omega(\tilde{P}) = \tilde{\Omega}$ such that $d_1(w_1) \succ d_2(w_1) \succ d_2(w_2) \succ d_1(w_2)$, we have $d_1 \notin f(P)$ and $d_2 \in f(P)$. Analogously for (II). Suppose that (III) obtains. Consider $\Omega' = \{w_1, w_2\}$ and $P' = \{d_1, d_2, d_3\}$ with $\Omega(P') = \Omega'$. Suppose that $d_1(w_1) \succ d_3(w_1) \succ d_3(w_2) \succ d_2(w_1) \succ d_2(w_2) \succ d_1(w_2)$.[9] By properties (1) and (2), $d_1, d_2, d_3 \in f(P')$. By d_2 dominates d_1. By property (6), therefore, $d_1 \notin f(P')$, a contradiction. Thus case (III) is impossible. We claim that if (I) holds, f is the lexicographic maximin, and if (II), the lexicographic maximax. We shall assume for the duration of the proof that (I) holds. The argument is entirely analogous for (II). Choose $u \in \mathcal{U}$. Consider $\Omega = \{w_1, \ldots, w_n\}$ and $D_\Omega = \{d \mid \text{domain } d = \Omega\}$. Let $\succsim^*$ be the order on D_Ω induced by f. Choose $d_1, d_2 \in D$. Let σ_1, σ_2 be a permutations of Ω for which $\exists m(\leqslant n)$, such that

$$\forall i > m \quad u(d_1(w_{\sigma_1(i)})) = u(d_2(w_{\sigma_2(i)})) ,$$

and

$$\forall i, j \leqslant m \quad u(d_1(w_{\sigma_1(i)})) \neq u(d_2(w_{\sigma_2(j)})) .$$

If $m = 0, d_1 \sim {}^*d_2$ by property (11). Since this case is trivial, we shall assume that $m \geqslant 1$. By property (14), $d_1 \succsim {}^*d_2$ implies that $d_1^1 \succsim {}_1^* d_2^1$ where d_1^1, $d_2^1 \in D_{\Omega_1}, \Omega_1 = \{w_1, \ldots, w_m\}, d_k^1(w_i) = d_k(w_{\sigma_k(i)})$ for all $i \in \{1, \ldots, m\}$ and $k \in \{1, 2\}$ and $\succsim {}_1^*$ is the ordering on D_{Ω_1} induced by f. For $k \in \{1, 2\}$, let $M_k = \max_{w \in \Omega_1} u(d_k^1(w))$ and $m_k = \min_{w \in \Omega_1} u(d_k^1(w))$. Choose $\bar{w}, \bar{\bar{w}} \in \Omega_1$ such that $u(d_1^1(\bar{w})) = M_1$ and $u(d_2^1(\bar{\bar{w}})) = m_2$.

$$\text{Define } d_1^2(w) = \begin{cases} d_1^1(\bar{w}), \text{ if } u(d_1^1(w)) \neq m_1 \\ d_1^1(w), \text{ if } u(d_1^1(w)) = m_1 . \end{cases}$$

$$d_2^2(w) = \begin{cases} d_2^1(\bar{\bar{w}}), \text{ if } u(d_2^1(w)) \neq M_2 \\ d_2^1(w), \text{ if } u(d_2^1(w)) = M_2 . \end{cases}$$

By property (6), $d_1^1 \succsim {}_1^* d_2^1$ implies that $d_1^2 \succsim {}_1^* d_2^2$.

Let $H = \{(M_1, M_2), (M_2, m_2), (m_1, M_2), (m_1, m_2)\}$. We shall assume that the four elements of H are distinct (the other cases can be argued almost identically).

Let $\Omega_3 = \{w_1, w_2, w_3, w_4\}$ and define $d_1^3, d_2^3 \in D_{\Omega_3}$

where $\quad u(d_1^3(w_1)) = u(d_1^3(w_2)) = M_1$,

$$u(d_2^3(w_1)) = u(d_2^3(w_3)) = M_2 ,$$

$$u(d_1^3(w_3)) = u(d_1^3(w_4)) = m_1 ,$$

$$u(d_2^3(w_2)) = u(d_2^3(w_4)) = m_2 .$$

Let $\succsim {}_3^*$ be the ordering on D_{Ω_3} induced by f. By properties (6) and (13), $d_1^2 \succsim {}_1^* d_2^2$ implies that $d_1^3 \succsim {}_3^* d_2^3$. Choose $d_3^3 \in D_{\Omega_3}$ such that $u(d_3^3(w_1)) = u(d_3^3(w_2)) = M_2, u(d_3^3(w_3)) = u(d_3^3(w_4)) = m_2$. By property (11) $d_2^3 \sim {}_3^* d_3^3$. Therefore $d_1^2 \succsim {}_1^* d_2^2$ implies that $d_1^3 \succsim {}_3^* d_3^3$. Take $\Omega_4 = \{w_1, w_2\}$ and d_1^4, $d_2^4 \in D_{\Omega_4}$ with $u(d_k^4(w_1)) = M_k$ and $u(d_k^4(w_2)) = m_k$ for $k \in \{1, 2\}$. Let $\succsim {}_4^*$ be the ordering of D_{Ω_4} induced by f. Then $d_1^3 \succsim {}_3^* d_3^3$ implies that $d_1^4 \succsim {}_4^* d_2^4$. Collapsing the chain of implications, we obtain

$$d_1 \succsim {}^*d_2 \text{ implies } d_1^4 \succsim {}^4 d_2^4 .$$

By construction $M_1 \neq M_2, M_1 \neq m_2, m_1 \neq M_2, m_1 \neq m_2$.

From (I) and property (6), $d_1^4 \gtrsim {}_4^* d_2^4$ implies that one of the following relations hold

(1) $M_1 > m_1 > M_2 > m_2$.

(2) $M_1 > m_1 > M_2 = m_2$.

(3) $M_1 = m_1 > M_2 > m_2$.

(4) $M_1 = m_1 > M_2 = m_2$.

(5) $M_1 > M_2 > m_1 > m_2$.

(6) $M_1 > M_2 = m_2 > m_1$.

(7) $M_2 > M_1 > m_1 > m_2$.

(8) $M_2 > M_1 = m_1 > m_2$.

Suppose that both $d_1^4 \gtrsim {}_4^* d_2^4$ and (6) hold. Choose M_2' and $m_2' \in \mathbb{R}$ and $d_3^4 \in D_{\Omega_4}$ such that $m_2 > M_2' > m_2' > m_1$, $u(d_3^4(w_1)) = M_2'$, and $u(d_3^4(w_2)) = m_2'$.[10]

By property (6), $d_2^4 \succ {}_4^* d_3^4$. Therefore $d_1^4 \succ {}_4^* d_3^4$. But, by (I), $d_3^4 \gtrsim {}_4^* d_1^4$; a contradiction. Therefore, if $d_1^4 \gtrsim {}_4^* d_2^4$, (6) is impossible. Let $\gtrsim {}_L^*$ be the ordering induced on D_Ω by the lexicographic maximin criterion.

Observe that any of the cases (1)–(5), (7) and (8) are consistent with $d_1 \gtrsim {}_L^* d_2$. Therefore, $d_1 \gtrsim {}^* d_2$ implies that $d_1 \gtrsim {}_L^* d_2$. By repeating essentially the same argument, we can show that $d_1 \succ {}^* d_2$ implies that $d_1 \succ {}_L^* d_2$. Therefore f is the lexicographic maximin.

Q.E.D.

The following result is an immediate corollary.

THEOREM 14. A criterion f which satisfies properties (1), (2), (6), (11), (13), (14), (15), (16) is the lexicographic maximin.

Proof. By Theorem 9, f is either the lexicographic maximin or maximax. By (16), f cannot be the latter.

The final result of this part is a new characterization of the ordinary maximin criterion. Its principal advantages over the axiomatizations in Theorems 4 and 9 are (i) it does not require a convexity axiom and (ii) it does not depend on a denseness assumption, nor, indeed, on any other assumption about the indifference classes of $\gtrsim$.

THEOREM 15. A criterion f which satisfies axioms (1), (2), (5), (10), (11), (12), (15), (16) is the maximin criterion.

Proof. Choose $u \in \mathcal{U}$. Properties (1), (2), (5) and (10) together imply that (4) holds. Thus, the hypotheses of Theorem 1 are satisfied, and there exists a weak ordering $\succsim_u^*$ as in the statement of Theorem 1. Choose $P \in \mathcal{P}$, d, $d' \in P$ as in property (16). Let $m = \min_w u(d(w))$, $m' = \min u(d'(w))$, $M = \max u(d(w))$, $M = \max u(d'(w))$. Then $M < M'$. By property (4), $m > m'$. So,

$$(9) \quad m' < m \leqslant M < M'.$$

Suppose there exists Ω such that for some $d_1, d_2 \in D_\Omega$, $d_1 \succsim_u^* d_2$ but $m_1 < m_2$ where $m_1 = \min(d_1(w))$ and $m_2 = \min(d_2(w))$. By (5), we have $m_1 < m_2 \leqslant M_2 \leqslant M_1$ where $M_1 = \max u(d_1(w))$, $M_2 = \max u(d_2(w))$. Choose $d_1' \in D_\Omega$ such that $m_1 < m_1' < m_2 \leqslant M_2 \leqslant M_1 < M_1'$.[11] Then, by (5), $d_1' \succ_u^* d_1 \succsim_u^* d_2$. But, by (15) and (9), $d_2 \succ_u^* d_1$, a contradiction. Therefore, $\forall \Omega \, \forall d_1, d_2 \in D_\Omega$, $\min u(d_1(w)) > \min u(d_2(w))$ implies that $d_1 \succ_u^* d_2$. By continuity (property 10), $\min u(d_1(w)) \geqslant \min u(d_2(w))$ implies that $d_1 \succsim_u^* d_2$. Therefore, f is the maximin criterion.

Massachusetts Institute of Technology

SUMMARY OF THE PRINCIPAL PROPERTIES

Property	
(1) (2)	The ranking of decisions must be an ordering
(4)	Weak Dominance
(5)	Strict Dominance
(6)	Admissibility
(7)	Column Linearity
(8)	Linearity
(9) (9')	Convexity
(10)	Continuity
(11)	Symmetry
(12)	Deletion of Repetitious States
(13)	Weakened Form of Deletion of Repetitious States
(14)	Independence of Non-discriminating States
(15)	Ordinality
(16)	Weak Pessimism

SUMMARY OF RESULTS

WITH THE VON NEUMANN-MORGENSTERN AXIOMS

Properties	H	M (Milnor)	IR (Chernoff)	IR (Milnor)	IR (Maskin)
(1)	⊗	⊗	⊗	⊗	⊗
(2)	⊗	⊗	X	⊗	⊗
(3)	X	X	⊗	X	X
(4)	X	X	⊗	X	X
(5)	X	X	X	⊗	X
(6)			⊗	X	X
(7)			⊗	⊗	X
(8)	⊗	X	X	X	⊗
(9)		⊗	⊗	X	X
(10)	⊗	⊗	X	X	⊗
(11)	⊗	⊗	⊗	⊗	⊗
(12)	⊗	⊗			
(13)	X	X			
(14)			X	X	⊗
(15)		X			
(16)	X *	X	X	X	X

Key: H – Hurwicz criterion   X – Criterion satisfies this property
        M – Maximin criterion   ⊗ – Criterion axiomatized by this property
        LM – Lexicographic maximin criterion   * – satisfies this property only if $\alpha \neq 1$.
        IR – Principle of Insufficient Reason

WITHOUT THE VON NEWMANN-MORGENSTERN AXIOMS

Properties	M (Modified Milnor)	LM	LM	M (Maskin)
(1)	⊗	⊗	⊗	⊗
(2)	⊗	⊗	⊗	⊗
(3)	X	X	X	X
(4)	X	X	X	X
(5)	X	X	X	⊗
(6)		⊗	⊗	
(7)				
(8)	X	X	X	X
(9')	⊗	X	X	X
(10)	⊗			⊗
(11)	⊗	⊗	⊗	⊗
(12)	⊗			⊗
(13)	X	X	⊗	X
(14)		⊗	⊗	
(15)	X	⊗	⊗	⊗
(16)	X	⊗	⊗	⊗
Denseness Assumption	⊗			
Countability Assumption	X	⊗		

NOTES

[1] A lottery is finite if it has only finitely many branches.

[2] Finiteness is a convenient but unnecessary supposition.

[3] '$x \sim y$' denotes '$x \succsim y$ and $y \succsim x$'.

[4] This is convenient for notational purposes but substantively unnecessary.

[5] We may assume that Ω contains at least three elements because if not, we can always add extra states via property (12).

[6] A mapping $g : P \to P'$ is $\succsim$-preserving if $\forall d_1, d_2 \in P \, \forall \, w, w' \in \Omega(P)$, $d_1(w) \succsim d_2(w')$ implies that $g(d_1)(w) \succsim g(d_2)(w')$ and $d_1(w) \succ d_2(w')$ implies that $g(d_1)(w) \succ g(d_2)(w')$.

[7] The assumption that $\mathcal{V}$ induces all possible orderings of $X \times N$ is actually equivalent to Hammond's Unrestricted Domain Condition [12].

[8] These exist by our assumption about the number of indifference classes of $\succsim$.

[9] See Note 8.

[10] M_2' and m_2' will not exist if M_1, M_2, and m_1 represent adjacent utility levels. However, by our assumption that there are at least six indifference classes and by property (15), we may as well assume that there is a gap of at least two indifferences classes between M_2 and m_1, so that M_2' and m_2' will exist.

[11] d_1' may not actually exist for reasons similar to those of Note 10. However, we may assume its existence, without loss of generality, by the same argument as above.

BIBLIOGRAPHY

[1] Arrow, K. J.: 1963, Social Choice and Individual Values (Wiley, New York).

[2] Arrow, K. J. and Hurwicz, L.: 1972, 'An optimality criterion for decision-making under ignorance', in D. F. Carter and F. Ford (eds.), Uncertainty and Expectations in Economics (Oxford).

[3] Chernoff, H.: 1954, 'Rational selection of decision functions', Econometrica 22, 422–443.

[4] d'Aspremont, C. and Gevers, L.: 1975, 'Equity and the informational basis of collective choice', presented at the Third World Econometric Congress, mimeographed.

[5] Debreu, G.: 'Topological methods in cardinal utility theory', in Arrow, Karlin, and Suppes (eds.), Mathematical Methods in the Social Sciences (Stanford University Press, Stanford).

[6] Herzberger, H.: 1973, 'Ordinal choice structures', Econometrica, pp. 187–238.

[7] Luce, R. D. and Raiffa, H.: 1957, Games and Decisions (Wiley, New York).

[8] Maskin, E. S.: 1975, 'A theorem on utilitarianism', mimeographed, Cambridge University.

[9] Milnor, J.: 1954, 'Games against nature', in Thrall, Coombs, and Davis (eds.), Decision Processes (Wiley, New York).

[10] Sen, A. K.: 1970, Collective Choice and Social Welfare (Holden Day, San Francisco).

[11] Strasnick, S.: 'Arrow's paradox and beyond', mimeographed (Harvard University).

[12] Hammond, P. J.: 1975, 'Equity, Arrow's conditions, and Rawls's difference principle', mimeographed, forthcoming in: Econometrica.